Astrophysics and Space Science Proceedings

For further volumes:
http://www.springer.com/series/7395

Environment and the Formation of Galaxies: 30 years later

Proceedings of Symposium 2 of JENAM 2010

Editors

Ignacio Ferreras
Anna Pasquali

Springer

SOCIEDADE PORTUGUESA ASTRONOMIA

Editors
Ignacio Ferreras
University College London
Mullard Space Science Laboratory
Holmbury St. Mary
RH5 6NT Dorking Surrey
United Kingdom
ferreras@star.ucl.ac.uk

Anna Pasquali
Astronomisches Rechen-Institut
Moenchhofstrasse 12-14
69120 Heidelberg
Germany
pasquali@ari.uni-heidelberg.de

ISSN 1570-6591 e-ISSN 1570-6605
ISBN 978-3-642-26995-0 ISBN 978-3-642-20285-8 (eBook)
DOI 10.1007/978-3-642-20285-8
Springer Heidelberg Dordrecht London New York

Cover design: eStudio Calamar S.L.

Printed on acid-free paper

Springer is part of Springer Science+Business Media (www.springer.com)

Preface

The publication of the *morphology–density* relation by Alan Dressler in 1980 brought into the limelight the role played by environment in the formation and evolution of galaxies. During the following three decades, we have learned that galaxy evolution is driven by both intrinsic processes (typically controlled by galaxy mass, also known as *nature*) and environmental effects (usually referred to as *nurture*). At present, we still have to understand the details of the interplay between nature and nurture. The advent of large, homogeneous redshift surveys has been a major step forward in this direction, since it has allowed us to quantify environment in a more consistent way, sampling a large variety of galaxy environments (from voids to massive galaxy clusters, through different size galaxy groups). Large galaxy surveys at different wavelengths have enabled us to study how different galaxy properties (e.g. morphology, star formation, stellar populations, AGN activity) depend on environment. The comparison between the observations and the predictions from state-of-the-art semi-analytical models and numerical simulations of galaxy formation in a cosmological context has proven essential in disentangling the mass assembly history from the star formation history of galaxies, in connection with the environment where they live.

The year 2010 represents the 30th anniversary of the *morphology–density* relation, and we took this opportunity to organize the symposium *Environment and the Formation of Galaxies: 30 years later*, with the purpose of establishing the impact of environment on the evolution of galaxies and its dependence on look-back time. Special emphasis was given to the physical mechanisms that are responsible for transforming galaxies once they are accreted by a group or a cluster (e.g. ram pressure and tidal stripping), including the observable imprint left in the galaxy HI distribution. Other major topics of the symposium were the environmental dependence of galaxy properties at $z \geq 1$ and the implementation of environmental effects in cosmological models of galaxy formation and evolution.

This symposium was hosted by the 2010 Joint European and National Astronomy Meeting, held in Lisbon on 6–10 September 2010. We thank André Moitinho de Almeida and the 2010 JENAM Scientific and Local Organizing Committees for their support and help with the logistics of the symposium.

We warmly thank the other members of the Scientific Organizing Committee of this symposium: Alan Dressler (Carnegie Observatories), Gabriella De Lucia (INAF – Trieste), Sadegh Khochfar (MPE Garching), Bianca Maria Poggianti (INAF – Padova) and Frank van den Bosch (Yale University), for their priceless help in planning such an interesting symposium, and all the speakers and participants for making it successful.

Last, but not least, we would like to gratefully acknowledge financial support from The Royal Astronomical Society of the UK, and the Max Planck Institut für Astronomie (Heidelberg, Germany). Their support has been essential for the organisation of this symposium.

Heidelberg and London *Anna Pasquali*
January 2011 *Ignacio Ferreras*

Contents

Contributors

Eduardo Amores SIM–Lisboa, Faculdade de Ciencias da, Universidade de Lisboa, Lisbon, Portugal, amores@sim.ul.pt

M.A. Aragón-Calvo The Johns Hopkins University, 3701 San Martin Drive, Baltimore, MD 21218, USA

Alfonso Aragón-Salamanca School of Physics and Astronomy, University of Nottingham, Nottingham NG7 2RD, UK, alfonso.aragon@nottingham.ac.uk

Kateřina Bartošková Faculty of Science, Masaryk University, Brno, Czech Republic
and
Astronomical Institute, Academy of Sciences of the Czech Republic, Prague, Czech Republic, katka.bartoskova@gmail.com

Verena Baumgartner Institut f. Astronomie, Universitat Wien, Turkenschanzstrae 17, 1180 Vienna, Austria, verena.baumgartner@univie.ac.at

B. Beygu Kapteyn Astronomical Institute, Univ. Groningen, Groningen, the Netherlands

Asmus Böhm Institute of Astro- and Particle Physics, University of Innsbruck, Technikerstrasse 25/8, 6020 Innsbruck, Austria, Asmus.Boehm@uibk.ac.at

Samuel Boissier Laboratoire dAstrophysique de Marseille, UMR6110, CNRS/ Universite de Provence, 38, rue Frdric Joliot-Curie, 13388 Marseille cedex 13, France, boissier@oamp.fr

Fernando Buitrago School of Physics and Astronomy, University of Nottingham NG7 2RD, UK, ppxf@nottingham.ac.uk

Kevin Casteels Departament dAstronomia i Meteorologia, Universitat de Barcelona, Mart Franques 1, 08028 Barcelona, Spain, kcasteels@am.ub.es

Ana L. Chies-Santos Astronomical Institute Utrecht, Princetonplein 5, 3584, Utrecht, The Netherlands, A.L.Chies@uu.nl
Bruno Coelho Centro de Astrofisica da, Universidade do Porto, Rua das estrelas 4150-762 Porto, Portugal
and
Faculdade de Ciencias, Universidade do Porto, Rua do Campo Alegre 4169-007 Porto, Portugal, up000308008@alunos.fc.up.pt

Arianna Cortesi School of Physics and Astronomy, University of Nottingham, University Park, Nottingham NG7 2RD, UK, acortesi@eso.org

Olga Cucciati Laboratoire dAstrophysique de Marseille, UMR 6110 CNRS-Universite de Provence, 38 rue Frederic Joliot-Curie, 13388 Marseille Cedex 13, France, olga.cucciati@oamp.fr

Gabriella De Lucia INAF Astronomical Observatory of Trieste, via G. B. Tiepolo 11, 34143 Trieste, Italy, delucia@oats.inaf.it

Alan Dressler Carnegie Observatories, 813 Santa Barbara Street, Pasadena, CA 91101, USA, dressler@ociw.edu

Ivana Ebrová Astronomical Institute, Academy of Sciences of the Czech Republic, Prague, Czech Republic
and
Faculty of Mathematics and Physics, Charles University, Prague, Czech Republic, ivana@ig.cas.cz

Paul Eigenthaler Institut für Astronomie, Universität Wien, Türkenschanzstrae 17, 1180 Vienna, Austria, paul.eigenthaler@univie.ac.at

M. Carmen Eliche-Moral Departamento de Astrofísica, Universidad Complutense de Madrid, 28040, Madrid, Spain, mceliche@fis.ucm.es

Anna Ferré-Mateu Instituto de Astrofsica de Canarias, 38205 La Laguna, Spain, aferre@iac.es

Ignacio Ferreras MSSL, University College London, Holmbury St Mary, Dorking, Surrey RH5 6NT, UK, ferreras@star.ucl.ac.uk

J.H. van Gorkom Dept. of Astronomy, Columbia University, New York, NY 10027, USA

Marco Grossi CAAUL, Observatorio Astronomico de Lisboa, Universidade de Lisboa, Lisbon, Portugal, grossi@oal.ul.pt

Ruth Grützbauch University of Nottingham, Nottingham, UK, ruth.grutzbauch@nottingham.ac.uk

Boris Häußler School of Physics and Astronomy, University of Nottingham, University Park, Nottingham NG7 2RD, UK, Boris.Haeussler@nottingham.ac.uk

Marc Huertas-Company Paris Observatory, 5 Place Jules Janssen, 92195 Meudon, France, marc.huertas@obspm.fr

Yara L. Jaffé School of Physics and Astronomy, University of Nottingham, University Park, Nottingham NG7 2RD, UK, ppxyj@nottingham.ac.uk

T. Jarrett IPAC, Spitzer Science Center, JPL, Caltech, Pasadena, CA 91125, USA

Eva Jütte Astronomisches Institut der RuhrUniversitat Bochum, Bochum, Germany, eva.juette@astro.rub.de

Sadegh Khochfar Max-Planck-Institute for Extraterrestrial Physics, Giessenbachstrasse 1, 85748 Garching, Germany, khochfar@mpe.de

Alexei Kniazev SAAO, P.O. Box 9, 7935 Observatory, Cape Town, South Africa, akniazev@saao.ac.za

K. Kovač Max Max-Planck-Institut für Astrophysik, D-85748 Garching, Germany

K. Kreckel Dept. of Astronomy, Columbia University, New York, NY 10027, USA

Francesco La Barbera INAF-OAC, Salita Moiariello 16, 80131, Naples, Italy, labarber@na.astro.it

Ewa L. Łokas Nicolaus Copernicus Astronomical Center, 00-716 Warsaw, Poland lokas@camk.edu.pl

Gary A. Mamon Institut d'Astrophysique de Paris, Paris, France, gam@iap.fr

Amata Mercurio Istituto Nazionale di Astrofisica, Osservatorio Astronomico di Napoli, Naples, Italy, mercurio@na.astro.it

Hugo Messias Centro de Astronomia e Astrofsica da Universidade de Lisboa, Observat orio Astronomico de Lisboa, Tapada da Ajuda, 1349-018 Lisbon, Portugal, hmessias@oal.ul.pt

Nigel L. Mitchell Institute für Astronomie, University of Vienna, Türkenschanzstrasse, 1180 Vienna, Austria, nigel.mitchell@univie.ac.at

Juan Carlos Muñoz-Cuartas Astrophysikalisches Institut Potsdam, An der Sternwarte 16, 14184 Potsdam, Germany, jcmunoz@aip.de

Anna Pasquali Astronomisches Rechen-Institut, Zentrum für Astronomie der Universität Heidelberg, Mönchhofstrasse 12–14, 69120 Heidelberg, Germany, pasquali@ari.uni-heidelberg,de

P.J.E. Peebles Joseph Henry Laboratories, Princeton University, Princeton, NJ 08544, USA

Salomé Pereira de Matos Observatório Astronómico de Lisboa, Centro de Astronomia e Astrofísica da Universidade de Lisboa, Tapada da Ajuda, 1349 019 Lisbon, Portugal, smatos@oal.ul.pt

Vasiliki Petropoulou Instituto de Astrofísica de Andalucía–CSIC, Glorieta de la Astronomía, 18008, Granada, Spain, vasiliki@iaa.es

Daniel J. Pisano Department of Physics, West Virginia University, P.O. Box 6315, Morgantown, WV 26506, USA, djpisano@mail.wvu.edu

E. Platen Kapteyn Astronomical Institute, Univ. Groningen, Groningen, the Netherlands

G. Rhee Dept. Physics & Astronomy, UNLV, Las Vegas, NV 89154-4002, USA

Luigi Secco Department of Astronomy, vic. Osservatorio 3, 35122 Padova, Italy, luigi.secco@unipd.it

David Sobral Institute for Astronomy, ROE, Blackford Hill, Edinburgh, UK, drss@roe.ac.uk

Abiy G. Tekola South African Astronomical Observatory, P.O. Box 9, Observatory 7935, Cape Town, South Africa, abiy@saao.ac.za

Daniel Thomas Institute of Cosmology and Gravitation, University of Portsmouth, Dennis Sciama Building, Burnaby Road, Portsmouth PO1 3FX, UK, daniel.thomas@port.ac.uk

Sheona Urquhart University of Victoria, Victoria, BC, Canada, surquhar@uvic.ca

Rien van de Weygaert Kapteyn Astronomical Institute, University of Groningen, P.O. Box 800, 9700AV Groningen, The Netherlands, weygaert@astro.rug.nl

Frank van den Bosch Astronomy Department, Yale University, P.O. Box 208101, New Haven, CT 06520-8101, USA, vdbosch@yale.edu

J.M. van der Hulst Kapteyn Astronomical Institute, Univ. Groningen, Groningen, the Netherlands

Simone Weinmann Leiden Observatory, Leiden University, P.O. Box 9513, 2300 RA Leiden, The Netherlands, weinmann@strw.leidenuniv.nl

Eric M. Wilcots University of Wisconsin, 475 North Charter St., Madison, WI 53706, USA, ewilcots@astro.wisc.edu

David J. Wilman Max-Planck-Institut für extraterrestrische Physik, Giessenbachstrae, 85748 Garching, Germany, dwilman@mpe.mpg.de

C.-W. Yip The Johns Hopkins University, 3701 San Martin Drive, Baltimore, MD 21218, USA

The Morphology–Density Relationship: Looking Back, Thinking Back

A. Dressler

Abstract The work I did in the late 1970s leading to the morphology–density relation was done in a time of rising interest in how galaxies acquired different morphological types. I describe briefly here how I contributed to this effort by adding a large number of morphologies for galaxies in rich clusters and the field. The strong correlation that I discovered between galaxy type and local galaxy density ran counter to ideas at the time that emphasized processes tied to the global cluster environment. Instead, it provided some of the first evidence for a hierarchical picture – one in which the density of the environment into which a galaxy was born would be its lifetime legacy. Though often cited as a relation between galaxy morphology and the influence of present-epoch environment, the morphology–density relation was interpreted by me, from the first, as the influence of the early environment of galaxy formation, passed down by the hierarchical growth of structure. In fact, it seems increasingly likely that the more fundamental correlation of galaxy morphology is with galaxy mass, and that the morphology–density relation is basically an expression of the prevalence of more massive galaxies in regions of higher galaxy density.

1 Destiny

In the film "Back to the Future" a hyper-nervous George McFly is pushed by his time-traveling son to invite Lorraine – Marty's future mother – to the *Enchantment Under the Sea* dance – a crucial event in Marty's own existence which his visit to the past has apparently endangered. A nerd of the first magnitude, George predictably botches the "you are my destiny" line his Cyrano-playing son has written for him – "you are my density," he flusters.

A. Dressler (✉)
Carnegie Observatories, 813 Santa Barbara Street, Pasadena, CA 91101, USA
e-mail: dressler@obs.carnegiescience.edu

I. Ferreras and A. Pasquali (eds.) *Environment and the Formation of Galaxies: 30 years later*, Astrophysics and Space Science Proceedings,
DOI 10.1007/978-3-642-20285-8_1, © Springer-Verlag Berlin Heidelberg 2011

As it turns out, density is destiny – in a hierarchical universe. We recognize today that the environment into which a galaxy is born is the one it will experience for the rest of its life. A galaxy residing in one of today's rich clusters is likely to have been bumping elbows with its siblings since birth, while one out in the boonies has grown accustomed to the lack of company. For most of the attendees of this symposium, these are common ideas – younger astronomers have grown up within the cold-dark-matter paradigm that has cemented such notions. But, around 1980, when the first steps were taken to integrate galaxy formation into a cosmological context, this was certainly not the common picture. On the contrary, astrophysicists believed that galaxy formation was a relatively rapid process that happened before large-scale structure had a chance to develop. A forming galaxy, it was said, could not "know" whether its *destiny* was to live in a cluster, a group, or a void. This was the context within which my discovery of a remarkably good correlation of galaxy morphology with the present-day *local* density of its surroundings was surprising, if not suspicious. What seems natural today was hard to fit into the work that was shaping the field, at a time when the first real progress was being made in this important subject.

In reading the abstracts submitted for this JENAM symposium, *Environment and the Formation of Galaxies*, I was surprised how many claimed they would show that local density was not the real driver of galaxy evolution, as (they apparently thought) I had argued in my 1980 paper [6].

Of course, I am ridiculously grateful to the conveners of this Symposium, Anna Pasquali and Ignacio Ferreras, for recognizing the work I did three decades ago, but especially glad as well that I have a chance to come here and explain why the morphology–density relation ran counter to this field's prevailing winds. In fact, I interpreted what I had found as evidence for the importance of *early* envirornment in shaping a galaxy's morphology and was almost strident in arguing that the later environment where I had found these galaxies was not likely responsible.

Perhaps I should have called it "the morphology-destiny relation" – maybe that would have made clearer what I thought it was all about.

2 Galaxies and Their Environments: History on a Budget

References to a connection between different kinds of nebulae, presumably the extragalactic ones, and the crowdedness of their surroundings can be found back to the eighteenth and nineteenth centuries, in the work of William Herschel and son John. But, with the limited depth of visual observations, and confusion about what nebulae actually were, it was not until the 1930s when these impressions were formally acknowledged, at least in a qualitative sense. In 1936 Edwin Hubble wrote, in the "Realm of the Nebulae" [13]

> There are some indications of a correlation between characteristic type and compactness, the density of the cluster diminishing as the most frequent type advances along the sequence of classification.

and, somewhat less obtusely,

> ...dominance of late typed among isolated nebulae in the general field.

In their famous 1931 paper showing the correlation of distance and redshift, Hubble and Milton Humason [14] had taken note of the fact that, in the Virgo Cluster

> Nebulae of all types except irregular are represented, but elliptical nebulae and early spirals are relatively much more numerous than among nebulae at large.

In 1936 Clyde Tombaugh [29], discoverer of Pluto – once a planet – was the first to map the Pisces-Perseus supercluster, and he noted that, in the richest clusters of this vast structure, elliptical galaxies were more concentrated than spirals. Always with the eye for detail, Fritz Zwicky [32] reported in 1942 that S0 galaxies in the Virgo cluster are distributed like the elliptical galaxies, not like the spirals – a prelude to what was to come.

I want to emphasize that these early references to galaxy morphology and environment were not linked to ideas about how galaxies had formed and achieved different types – the common reference to Hubble's belief that the morphological sequence was one of galaxy age is a myth. The credit for motivating the field to think about galaxy origins goes, I think, to Walter Baade [2] for his work in the 1940s relating the different stellar populations of the Andromeda galaxy to the Milky Way's Population I and Population II stars – stellar populations with known histories. In 1962 Olin Eggen, Donald Lynden-Bell, and Allan Sandage [9] took Baade's work a step further when they proposed a link between metal abundance and kinematics in the Galaxy that argued for an early and orderly collapse phase of the gas cloud that became the Milky Way, a breakthrough that also led a decade later to the picture of a more chaotic collapse, by Leonard Searle and Bob Zinn [27]. Thinking along similar lines, but with a different approach, Morgan sought a connection of these earlier galaxy correlations to stellar population types in his 1962 Russell Lecture, by comparing galaxy types with old and young star clusters and emphasizing how these are distributed within galaxies (bulge, disk, and halo) [19]. In retrospect, perhaps the first morphology–density relation was Morgan's "discovery" of cD galaxies and his conclusion that they were only found in the dense centers of rich clusters.

These were great steps forward. Unfortunately, before the 1970s there was no clear way to connect either galaxy morphology or stellar populations to galaxy formation, that is, there was no cosmological model like ΛCDM to help explore how different types of galaxies might arise. Without a starting point for galaxy formation models, the only ideas that could reasonably be put forward and tested were things that could happen to galaxies later in their lives that might force a differentiation. For example, in 1951 Lyman Spitzer and Baade [28] proposed that galaxy collisions in clusters make S0 galaxies, stripping the disk gas as the stars in two galaxies pass through each other. The idea died as the Hubble constant fell, but, with the advent of X-ray astronomy and the remarkable discovery of hot dense gas in galaxy clusters, Jim Gunn and Richard Gott [12] revived the stripping model in 1972 by inserting the hot intracluster medium (ICM) as the agent of change. They predicted that ram pressure would strip the gas from infalling spirals as they plowed through the ICM of a dense cluster core. In the same year, successful modeling of strong tidal interactions of merging spirals by Alar and Juri Toomre [30] suggested a mechanism for the formation of elliptical galaxies – another process that could happen late in a galaxys life. A few more ideas followed for turning spiral galaxies into S0s, for example, Richard Larson et al. [16] suggested that the stripping of gas reservoirs in the cluster environment simply shut down star formation in a spiral, and Len Cowie and Toni Songaila [4] and Paul Nulsen [20] suggested that – in the dense environment of a cluster center – hot X-ray gas would "boil away" a spiral's intragalactic gas.

While these somewhat piecemeal attempts to explain the origin of galaxy morphologies were being advanced, a more substantial assault on the problem began. Jim Peebles and Marc Davis were interested in pushing back to the early universe to provide a physical context for galaxy formation and the growth of large scale structure. Ironically, this bold new step only reinforced the idea that galaxy differentiation into morphological types was a late-in-the-game process. Their framework for studying structure formation was what Tjeerd van Albada [1] had called *gravitational instability*, the simple idea that fluctuations in density in the early universe would irreversibly grow in amplitude as the universe expanded over cosmic time. The question was, how to describe these fluctuations at some early time, so that they could be evolved and compared with the present-day structure of the universe – the only real data available at the time. In 1977 Davis and Peebles [5] presented an analytical solution of the statistics of such an evolving gravitation system and compared it to the "counts in cells" of faint galaxies in the Shane-Wirtanen catalog – from observations done with the Lick Observatory astrograph and stored on *sheets of paper(!)* in Santa Cruz. The formalism used by Davis and Peebles was general, relying on the power-law distribution of the fluctuation amplitude at some early time. Based on statistics of the *correlation function*, they derived a power-law spectrum of initial fluctuations with an index of $\eta = 0$ – what is commonly called "white noise," in effect, equal power at all "wavelengths" of structure. In such a universe, smaller structures (galaxies and very small groups) formed and matured much earlier than larger ones, e.g., rich clusters. In this context, and assuming that galaxies began to coalesce early, differences among galaxies would logically be the result of later processes – galaxies, it was often said, could not "know" when they were forming where they were going to wind up. The pioneering work of Davis and Peebles seemed to seal the deal: galaxies "differentiated" later in life, or again – in the lingo of the day – it was "nurture, not nature."

3 Grad Student Seeks Exciting Career Opportunity

From 1970 to 1976 I was a graduate student at Lick Observatory, UC Santa Cruz. I had the very good fortune of being at a department that was "on the map" and so many of the principals in the studies I have just talked about passed through and gave colloquia. For my own part, I was working on a PhD thesis project that involved taking photographic plates of clusters of galaxies with the Lick Observatory Crossley telescope, and scanning them with a machine that had been modified by me for the purpose, using one of the very basic computers of the day, a Digital PDP-8. This photographic photometry served as the basis for a study of the luminosity functions (LFs) of a dozen clusters of galaxies, specifically aimed at the question of the universality of the cluster LF that George Abell had suggested. In the end, the differences I found were small, and the few cases that seemed to depart from the "universal" form were arguably due to the gobbling of other bright cluster galaxies by a central cD. The universality of cluster LFs was another piece of observational data that supported the idea that all galaxies had formed in more or less the same way – environment independent, but my evidence was not as strong as Paul Schechter's comparison of field and cluster LFs [25] (the work in which Schechter developed his irreplaceable parameterization of the LF). Schechter's finding that cluster galaxy and field galaxy LFs were the same, within the errors, was yet further evidence that galaxies were born in common conditions.

In hindsight, we all really missed something here, something that turned out to be relevant to the work I would do a few years later that led to the morphology–density relation. Yes, the luminosity functions of field and cluster galaxies were in close agreement, but luminosity is a time-dependent parameter that changes with aging stellar populations. A galaxy's mass, by comparison, is more fundamental, and not subject to the rapid changes that are possible in galaxy luminosities. Mass can change as well, as with major mergers, of course, but that activity has died down considerably in the last half of the Hubble time. By the end of the 1970s kinematic measurements were already being combined with accurate photometry to determine visible mass-to-light ratios (hereinafter M/L), and it was becoming clear that elliptical galaxies typically had M/L values 2–3 times higher than late-type spiral galaxies, as reviewed in 1979 by Faber and Gallagher [10]. Combining this with what was known about the difference in morphological mix of cluster and field, it should have been apparent that the *mass* functions of the field and cluster could not be the same if the luminosity functions were. But, I know of no astronomer, me included, who put two and two together, so to speak, and drew this important conclusion from Schechter's work. If this had been recognized, it would have thrown a curve into the straight line thinking that galaxies were born in ways that were independent of larger-scale environment, because – on average – clusters of galaxies contain more massive galaxies than the field. I will return to this subject and its relevance to the mophology-density relation in my closing.

4 Making the Morphology–Density Relation

Around 1973 Olin Wilson of the Mt. Wilson staff came to visit the astronomers at Santa Cruz. At an informal sit-around, he was asked about the new du Pont telescope that Carnegie was building at the Las Campanas Observatory. What was special about this new telescope, he explained, was the Ritchey-Cretien design that gave it a well-corrected 2-sq-deg field, intended for first use with 20-in. photographic plates. The plate scale would match the 200-in. Hale telescope prime focus, 11-arcsec per mm, but the du Pont would have a field size more like a Schmidt telescope. A couple of years later, when I applied for a Carnegie Fellowship, this came up in a discussion with my thesis advisor Sandy Faber, who, in helping me come up with a research proposal for the application, asked me what were the most important "missing" data for understanding clusters of galaxies. This was, of course, galaxy morphology – the telescopes that could photograph wide enough areas of sky to study clusters had too small a plate scale for robust classifications, while those with a suitable scale typically had small fields. The relatively new 4-m telescopes at Kitt Peak and Cerro Tololo offered the right combination, but more relevant for this conversation was the du Pont telescope that Wilson had described. Following the work of Augustus Oemler in his PhD thesis at Caltech, I proposed to make a more extensive and more accurate (because of the high plate scale) study of galaxy morphology and its relation to other properties of clusters and their galaxies. Thanks to Faber's insight and guidance, I was able to write a proposal that won me a Carnegie Fellowship, an opportunity I had not thought possible at the time.

The du Pont was not quite ready when I arrived at Carnegie in the fall of 1976, but by the following spring Horace Babcock – Director of the Hale Observatories and the driving force behind Las Campanas and the du Pont – had taken the first photographic plates and demonstrated the excellence and power of the new camera. Because other instrumentation for the new telescope was still months away, I was given a generous allotment – what amounted to two full dark runs – in August and October of 1977. Armed with hundreds of "finder charts" of clusters of galaxies, I became the first astronomer to begin a science program with the du Pont. The television guide cameras were not ready either, so I had to use the main eyepiece of the telescope to verify the fields and to guide by hand paddle, scrunched underneath the telescope on long, cold winter nights and bleary-eyed, and then half asleep trying to develop all the plates in the afternoons. The August run was almost completely clear and by the end I was a wreck – but that is another story.

Over the next 2 years I accumulated plates of some 40 clusters in seeing of 1 arcsec or better with the du Pont, and added another dozen or so clusters with plates I took at the Mayall telescope and the Palomar 60-in. telescope. Carnegie had a carpenter on the shop staff in those days and he built a large light box for me, copying a design he had made for Allan Sandage, who was also an early, intense user of the du Pont camera. I fitted the light box with a drafting machine that would allow me to scan the plates with an eyepiece, measure positions, and compare to images of galaxies and galaxy bulges for a rough magnitude scale – a technique that was called flyspanking. Although imprecise, the systematic errors from this approach seemed to be small enough to yield useful data; this work served as the basis for a later paper by Schechter and myself [26] where we estimated the global bulge-mass/disk-mass in the present-day universe to be about unity, a result I think has stood up all these years.

It took about 8 months, as I remember, to do a raster-scan by eye of all the plate material; I recorded the data by hand on accounting paper that was later transcribed to IBM computer cards, one per each of the ∼6,000 galaxies [7]. I then started to investigate the data, starting with a comparison to Gus Oemler's [21] result of three characteristic population ratios. In the 1980 paper there is a plot of where each of the clusters I studied falls on this plane of the percentage of ellipticals and percentage of spirals – these were global measures of the total population out to an "effective radius." I interpreted Oemler's result as three discrete population types, but Oemler later told me that the distribution was probably continuous and the three types were only meant to be representative. He was more right than he knew, unfortunately: I found only a spread of points all over the map, scratched my head and moved on.

I next looked at the work of Jorge Melnick and Wal Sargent [18] on the populations of six X-ray emitting clusters, for which they found a rising S0 fraction with increasing velocity dispersion (a global relation), and a rising spiral-to-S0 ratio as a function of radius, a semi-global relationship. Both were consistent, they claimed, with ram-pressure stripping of spirals to make S0 galaxies. However, the scale and slope of this relationship was sufficiently variable in their six-cluster sample that "consistent" was about all one could say – and the same could be said for the radial correlations for my eight clusters that were strong X-ray emitters.

I searched further for meaningful correlations, basically following the "plot anything against everything else" approach. This seldom works and it did not here. I was somewhat distressed by this point; it seemed that several years of very hard work might be leading nowhere. Almost in desperation, I tried the one thing that seemed ridiculous given the notions we all had about cluster formation. I had noticed, as I sloughed through the examination of each and every galaxy in my 55 clusters, what seemed to be discrete clumps of galaxies – they looked like real physical associations. From the dogma of the day, I also knew that this was not thought to be possible – I reasoned that perhaps these were just foreground and background groups superposed on my clusters. One reason for this shared prejudice was Donald Lynden-Bell's slick answer to the baffling question of how clusters of galaxies could have such steep radial density profiles [17], when two-body relaxation was far too slow, a paradox Zwicky had raised decades before [31]. Lynden-Bell described a process he named with the oxymoron "violent relaxation," in which massive galaxies exchanged kinetic energy with light ones through the intermediary of the violently changing gravitational potential of the collapsing cluster. This reproduced cluster profiles very well, but it also was sure to scramble galaxies in the cluster and break up any loose groups, as can be seen in the 1970 simulation of the violent relaxation of the Coma cluster by Peebles [22]. Then, too, there was the prevalent belief that galaxy morphology had to occur late and that "galaxies could not know where they were going to end up" idea. It just did not seem reasonable that there were distinct subgroups in rich clusters and that they had unique population mixes.

Nevertheless, clueless of what else to do, that is exactly what I decided to look into, writing a computer program that calculated the local projected density of each galaxy by measuring the rectangular area over which its ten nearest neighbors were

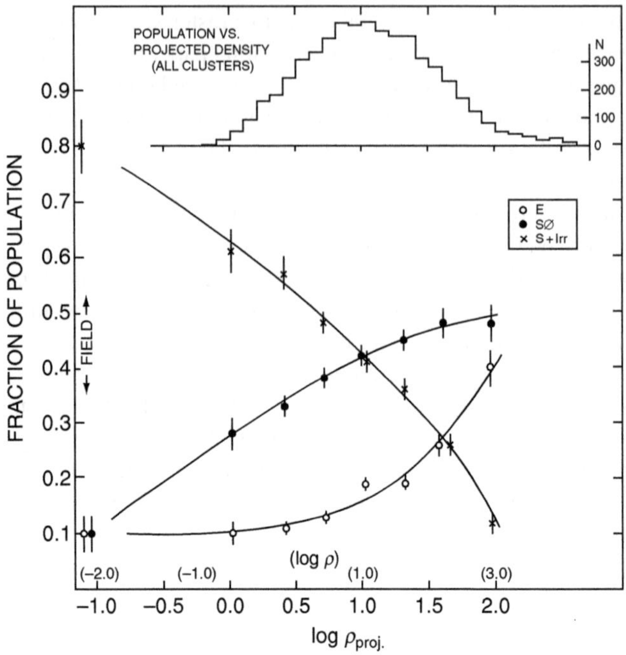

Fig. 1 The fraction of E, S0, and S+I galaxies as a function of the log of the projected density, in galaxies Mpc^{-2}. The data shown are for all cluster and field galaxies in the 1980 sample. Also shown is an estimated scale of true space density in galaxies Mpc^{-3}. The upper histogram shows the number distribution of the galaxies over the bins of projected density. Figure from [6]

distributed. The program ran on one of the central mainframe computers at Caltech, with the two boxes of punchcards that recorded the data. A few minutes of CPU time on the IBM mainframe computer of the day was required to do this millisecond's of work on my Powerbook. I ran the first half one night and plotted up the data. Barely able to contain my excitement and to resist telling anybody what I thought I had found, or even to sleep, I ran the second half over the following night.

What emerged was the now familiar diagram of Fig. 1. What I had found astonished me, especially given the weeks of casting about for some correlation good enough to think about. I had the greatest enjoyment and satisfaction in my astronomical career trying to figure out what this excellent correlation meant. Although this was early in my career, I somehow knew already that moments like this were rare – I wish that every scientist could have this good fortune. In short order I had compared the morphology–density relation to the morphology-radius relation for the same clusters and found, unsurprisingly, that in very regular clusters with little substructure the results were very similar, but that the morphhology-density relation held even in very clumpy, irregular clusters, where population correlated poorly with radius. Clearly, it seemed that the clumps were real physical associations and the correlation of morphology was with a local rather than global parameter. The implication was that galaxies had gotten their morphological type from some process that had more to do with their local environment than some global process such as ram-pressure stripping. I did look to see if I was missing any "second parameter" to the morphology–density relation, but the only one I found was that the strong X-ray emitting clusters seemed to have slightly offset curves, meaning fewer spirals and more E's and S0's at a given density – perhaps that was a second-order global effect.

I am still surprised that through these many years there have been occasional papers arguing that the morphology-radius relation is the fundamental relation, not the morphology–density relation. My own work aside, what I find decisive is the remarkable study by Riccardo Giovanelli et al. [11] of the Pisces-Perseus superclusters in which they show maps across a wide swath of sky sliced by morphological type. Figures 1 and 6 of that paper show a vivid correlation in pictorial form of galaxy morphology with local-density over a wide range of environments, from rich clusters, through rich groups, to poor groups, to isolated field galaxies. The rich clusters – where global effects might plausibly dominate – look to be part of smooth, continuous relationship, rather than a special environment. In their study of the *low-density group* regime of the morphology–density relation, Marc Postman and Margaret Geller [23] found the same behavior as did Giovanelli and collaborators. I still find the evidence from these studies to be compelling, but by now there is also the well accepted paradigm of the hierarchical growth of structure that seems to me to obviate global processes as the primary driver of morphological differentiation.

5 What Did It Mean? For That Matter, What *Does* It Mean?

It seems amazing to admit that even 30 years later we are still not sure why there is a strong correlation of galaxy type with local density. From the start, many in the field pointed to it as further evidence for mergers or ram-pressure stripping in the shaping of galaxy morphology. Qualitatively, of course, the disappearance of spirals in the dense regions of rich clusters looked like evidence for ram pressure stripping, or perhaps the Larson et al. stripping of gas reservoirs from spirals. But, a more careful look at the morphology–density relation showed that the S0 fraction rises more slowly than the spiral fraction plummets – no one then (or now) expected that some of those supposedly stripped spirals were adding to the elliptical population. Furthermore, the upturn in the elliptical fraction – coming as it does for the densest regions – is certainly not what is expected if late mergers were responsible, because the high velocity dispersions of such regions strongly discourage them. So, right from the start, I thought that the morphology–density relation was in conflict with these popular ideas.

For me there was one other point that was compelling for the question of late processes transmuting galaxy types. How could the fractions change so slowly, I wondered, over two or three orders of magnitude in space density? What kind of physical process was so weak and yet apparently so effective? I expressed the points above, and my answer to this last question, in the 1980 paper, arguing quite strongly (I thought) that the morphology density relation must be a *legacy* of an earlier time, when the density contrast was much weaker, before these structures grew to the high contrast we see today. I was careful, therefore, not to identify the present-epoch local density of these environments as the agent of change – it did not fit the evidence. Instead I considered the present-day environment to be a proxy for much earlier processes – early mergers in groups, for example, before the velocity dispersions grew large, and maybe some sort of stripping process that occurred in the much more dynamic, chaotic environment of the first billion years, before there were well-formed spiral galaxies.

This I thought was clear, but there *was* something in this picture that puzzled me. I was impressed by the Davis and Peebles "white noise" spectrum for the initial fluctuations, and this argued against galaxies experiencing an early differentiation of morphological "destiny," if you will, because the environment manifest itself too late in this picture. Later that year I was visiting my uncle, Robert Dressler – he had been my first mentor, a broadly knowledgeable physicist who had spent a good deal of his professional life understanding, developing, and applying signal processing techniques for radar applications during the cold war – I cannot say more. He was very excited about my new finding, but when I brought up my puzzlement about the Davis and Peebles initial power spectrum being at-odds with what I considered to be the implications of the morphology–density relation, he immediately cut me off: "Nothing in nature has a white noise spectrum," Uncle Bob insisted, "It's always 1/f noise."[1] If this were true for the fluctuations of the primordial density field, there would be more power at longer wavelengths relative to shorter wavelengths than for a white noise spectrum – at all times, as this system evolved by gravitational instability. I was quite taken with this idea, so for the next year, as I toured around, the talk I gave on the morphology–density relation included a transparency that showed how a 1/f spectrum with relatively more power at longer wavelengths could explain the correlations at early times with environment – because of this, I argued, galaxies *could* be influenced, as they were forming, by the early growth of the large-scale structure they would one day inhabit. Regretfully, I removed this key discussion from the 1980 paper – bad advice from a learned, more experienced colleague who thought the 1/f noise idea was too ad hoc. (See, you don't have to be an Einstein to have a "biggest blunder.") Only a couple of years later I heard about cold-dark-matter and its 1/f power spectrum, as that model took the field by storm. In 1986 Nick Kaiser [15] worked through the statistics of this coupling of scales and described the implications for galaxy formation and large-scale clustering.

Looking back, I was on the right track. All the focus on late-time processes that preceded my work became less interesting, I think, once there was a successful model for structure growth that allowed us to consider the correlations of galaxy development with their local environments over their lifetimes. As I said up front, in such a universe, density is destiny.

---

[1]Inverse frequency, also called "pink noise."

There are many things that have happened since 1980 that have made me more comfortable with the morphology–density relation – too much to go into here, but I would like to point to the discovery by the "Morphs" [8] that the morphology-relation has evolved since $z = 1$. In particular, the S0 population really does seem to have grown from the demise of the spiral population – it is more of a "late-time" development than I had imagined. That work showed that the elliptical galaxies are in place from early times (see also Postman et al. [24]). Perhaps, this is true of many of the S0 galaxies as well, but there is a dramatic increase in their numbers as the population of spirals – first discovered by Butcher and Oemler [3] in intermediate-redshift clusters – are cut down in their prime (see also the contribution here by Alfonso Arragon-Salamanca). I have come to believe that this is driven mostly by local interactions, tidal and mergers, and that ram-pressure stripping does play a substantial role at the end of the process, but the full picture is still far from clear and woefully lacking in detail. Hopefully, this will improve with the astonishing new capabilities of the new facilities that are under construction, for both ground and space.

I want to conclude by returning to my 1980 paper and highlighting this one point that – to my surprise – was apparently not appreciated. I never said that local density was the *agent* of change. In fact, as I have just described, I argued that the present-epoch local density I was measuring was probably *not* responsible for the differentiation of galaxies. Unfortunately, this was more of a negative conclusions, something that was *not* the cause, but for many years I have been promoting a more positive interpretation. As I said earlier in the discussion of the luminosity functions of field and cluster galaxies, the increase in mass-to-light ratio across the Hubble sequence from spirals to ellipticals suggests that the galaxies in dense regions are systematically more massive than those in the field. It seems quite reasonable to suggest, then, that the main driver (and predictor) of morpholgy is galaxy mass (baryonic or halo mass – they track together). In this picture, the morphology–density relation is merely an expression of this: higher density regions have systematically higher mass galaxies, and these are more likely to be early types, large-bulge disk galaxies and ellipticals. I leave you with that thought, that the morphology–density relation is but an echo of galaxies formation long ago, when density became destiny.

Acknowledgements I would like to acknowledge the efforts of all the members of the organizing committee and express my gratitude that you all have come here to Lisbon and acknowledged my work on the morphology–density relation – I feel very honored.

References

1. G.V. van Albada, AJ **66**, 590 (1961)
2. W. Baade, ApJ **100**, 137 (1944)
3. H. Butcher, A. Oemler Jr., ApJ **219**, 18 (1978)
4. L.L. Cowie, A. Songaila, Nature **266**, 501 (1977)

5. M. Davis, P.J.E. Peebles, ApJS **34**, 425 (1977)
6. A. Dressler, ApJ **236**, 351 (1980)
7. A. Dressler, ApJS **42**, 565 (1980)
8. A. Dressler, et al., ApJ **490**, 577 (1997)
9. O.J. Eggen, D. Lynden-Bell, A.R. Sandage, ApJ **136**, 748 (1962)
10. S.M. Faber, J.S. Gallagher, ARAA **17**, 135 (1979)
11. R. Giovanelli, M.P. Haynes, G.L. Chincarini, ApJ **300**, 77
12. J.E. Gunn, J.R. Gott, ApJ **176**, 1 (1972)
13. E. Hubble, *The Realm of the Nebulae* (Yale University Press, New Haven, 1936)
14. E. Hubble, M.L. Humason, ApJ **74**, 43 (1931)
15. N. Kaiser, in *Inner and Outer Space* (University of Chicago Press, Chicago, 1986), pp. 258–263
16. R.B. Larson, B.M. Tinsley, R.N. Caldwell, ApJ **237**, 692 (1980)
17. D. Lynden-Bell, MNRAS **136**, 101 (1967)
18. J. Melnick, W.L.W. Sargent, ApJ **215**, 401 (1977)
19. W.W. Morgan, ApJ **135**, 1 (1962)
20. P.E.J. Nulsen, MNRAS **198**, 1007 (1982)
21. A. Oemler Jr., ApJ **194**, 1 (1997)
22. P.J.E. Peebles, AJ **75**, 13 (1970)
23. M. Postman, M.J. Geller, ApJ **281**, 95 (1984)
24. M. Postman, et al., ApJ **623**, 721 (2005)
25. P.L. Schechter, ApJ **203**, 297 (1976)
26. P.L. Schechter, A. Dressler, ApJ **94**, 563 (1987)
27. L. Searle, R. Zinn, ApJ **225**, 357 (1978)
28. L. Spitzer Jr., W. Baade, ApJ **113**, 413 (1951)
29. C.W. Tombaugh, PASP **49**, 259 (1937)
30. A. Toomre, J. Toomre, ApJ **178**, 623 (1972)
31. I.F. Zwicky, PNAS **25**, 604 (1939)
32. I.F. Zwicky, ApJ **95**, 555 (1942)

The Cosmic Mass Density Field Reconstruction from the SDSS Group Catalog

J.C. Muñoz-Cuartas, V. Müller, and J.E. Forero-Romero

Abstract Based on the group catalog (GC) of the SDSS-DR4 [Yang et al., ApJ 671, 153 (2007)] we implemented a halo-based density field reconstruction method introduced by [Wang et al., MNRAS 394, 398 (2009)]. The method uses the mass distribution in and around dark matter structures computed from cosmological simulations to map the cosmic mass distribution. This approach enables us to make reconstructions of the density field avoiding the use of arbitrary smoothing functions and including information about the mass distribution in the environment of halos. We present the results of our reconstruction as well as the identification of different environments. The reconstructed density field will be further used in the investigation of environmental dependence of galaxy properties.

1 Description and Results

Following [2] we estimate the mass distribution in and around dark matter structures from a simulation of the spatially flat ΛCDM cosmology in a box of $100\,h^{-1}$Mpc and 512^3 particles run with a set of cosmological parameters compatible with those of WMAP3. To compute the average mass distribution in and around halos we binned them by mass in the range of [11.5,14.2] in $\log(M_{\rm vir}/h^{-1}M_{\odot})$ with each mass bin having a width of 0.3 dex. Figure 1 shows two of such distributions.

For the reconstruction we choose groups in the GC [3] with redshifts between 0.01 and 0.1, which offers an almost complete sample of halos with masses down to $10^{11.5}\,h^{-1}M_{\odot}$. Our final sample contains 88687 groups. Assuming the mass estimated for each dark matter halo in the catalog and the mass distribution we obtained from the simulations, we derive a realization of the density field from the

J.C. Muñoz-Cuartas (✉) · V. Müller · J.E. Forero-Romero
Astrophysikalisches Institut Potsdam, An der Sternwarte 16, 14184 Potsdam, Germany
e-mail: jcmunoz@aip.de

I. Ferreras and A. Pasquali (eds.) *Environment and the Formation of Galaxies: 30 years later*, Astrophysics and Space Science Proceedings,
DOI 10.1007/978-3-642-20285-8_2,

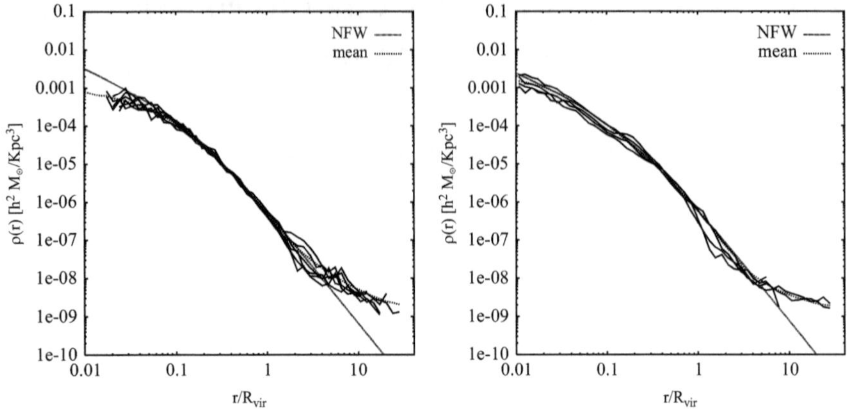

Fig. 1 Mass distribution obtained from the simulations for two different mass bins, $M = 10^{11.5}$ (*left*) and $M = 10^{13}\,h^{-1}M_{\odot}$(*right*). The small-dotted line shows the typical NFW density profile for that halo mass. The average density profile we estimate from the simulation is shown with the long-spaced doted line. The thin solid lines show some of the individual density profiles for each halo in that mass bin. Note the extension of the density profile out to 30 times R_{vir}, in the region defined as the domain of the halo

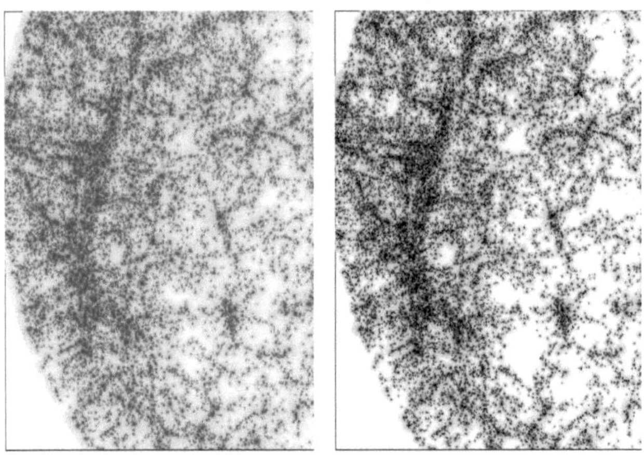

Fig. 2 (*Left*) A $300 \times 400\,h^{-1}$Mpc slice of the reconstruction for the north cap of the survey, where the structure of the great attractor can be clearly seen. (*Right*) Regions classified as peaks and filaments which clearly follows the distribution of galaxies in the large scale structure. It is possible to observe as well the empty regions (*white*) corresponding to voids in the density field

group catalog in redshift space. Our reconstruction consists on a set of $\sim 4 \times 10^7$ tracer *particles* of the density field following the mass distribution traced by the halos in the group catalog. For illustration, Fig. 2 shows a piece of the reconstructed density field.

Following [1], we classify a point in the density field as belonging to a void, sheet, filament or peak environment. Figure 2 (right) shows the identification of filaments and peaks (for a given threshold of the eigenvalues of the deformation tensor λ_{th}). To quantify the effects of the survey boundary and the influence of redshift distortions on our classification, we built a mock catalog from a cosmological simulation with a box size of $1\,h^{-1}$ Gpc and $1{,}024^3$ particles. We found that the volume fractions of the catalog representing the redshift space sample have been affected by a global factor of 20%, reflecting the increase in the mean density in non empty regions due to the effect of redshift distortions.

References

1. J.E. Forero-Romero et al., MNRAS **396**, 1815 (2009)
2. H. Wang et al., MNRAS **394**, 398 (2009)
3. X. Yang et al., ApJ **671**, 153 (2007)

The Void Galaxy Survey

R. van de Weygaert, K. Kreckel, E. Platen, B. Beygu, J.H. van Gorkom, J.M. van der Hulst, M.A. Aragón-Calvo, P.J.E. Peebles, T. Jarrett, G. Rhee, K. Kovač, and C.-W. Yip

Abstract The Void Galaxy Survey (VGS) is a multi-wavelength program to study ~60 void galaxies. Each has been selected from the deepest interior regions of identified voids in the SDSS redshift survey on the basis of a unique geometric technique, with no a prior selection of intrinsic properties of the void galaxies. The project intends to study in detail the gas content, star formation history and stellar content, as well as kinematics and dynamics of void galaxies and their companions in a broad sample of void environments. It involves the HI imaging of the gas distribution in each of the VGS galaxies. Amongst its most tantalizing findings is the possible evidence for cold gas accretion in some of the most interesting objects, amongst which are a polar ring galaxy and a filamentary configuration of void galaxies. Here we shortly describe the scope of the VGS and the results of the full analysis of the pilot sample of 15 void galaxies.

R. van de Weygaert (✉) · E. Platen · B. Beygu · J.M. van der Hulst
Kapteyn Astronomical Institute, Univ. Groningen, Groningen, the Netherlands

K. Kreckel · J.H. van Gorkom
Dept. of Astronomy, Columbia University, New York, NY 10027, USA

M.A. Aragón-Calvo · C.-W. Yip
The Johns Hopkins University, 3701 San Martin Drive, Baltimore, MD 21218, USA

P.J.E. Peebles
Joseph Henry Laboratories, Princeton University, Princeton, NJ 08544, USA

T. Jarrett
IPAC, Spitzer Science Center, JPL, Caltech, Pasadena, CA 91125, USA

G. Rhee
Dept. Physics & Astronomy, UNLV, Las Vegas, NV 89154-4002, USA

K. Kovač
Max Max-Planck-Institut für Astrophysik, D-85748 Garching, Germany

I. Ferreras and A. Pasquali (eds.) *Environment and the Formation of Galaxies: 30 years later*, Astrophysics and Space Science Proceedings,
DOI 10.1007/978-3-642-20285-8_3,

1 Introduction: Voids and Void Galaxies

Voids have been known as a feature of the Megaparsec Universe since the first galaxy redshift surveys were compiled [8, 16, 18]. *Voids* are enormous regions with sizes in the range of 20–50 h^{-1} Mpc that are practically devoid of any galaxy, usually roundish in shape and occupying the major share of space in the Universe [33]. Forming an essential ingredient of the *Cosmic Web* [4], they are surrounded by elongated filaments, sheetlike walls and dense compact clusters.

A major point of interest is that of the galaxies populating the voids. The pristine environment of voids represents an ideal and pure setting for the study of galaxy formation. Largely unaffected by the complexities and processes modifying galaxies in high-density environments, the isolated void regions must hold important clues to the formation and evolution of galaxies. This makes the relation between *void galaxies* and their surroundings an important aspect of the interest in environmental influences on galaxy formation [11, 14, 22, 24, 30].

Amongst the issues relevant for our understanding of galaxy and structure formation, void galaxies have posed several interesting riddles and questions. Of cosmological importance is the finding from optical and HI surveys that the density of faint galaxies in voids is only 1/100th that of the mean. As has been strongly emphasized by Peebles [23], this dearth of dwarf void galaxies cannot be straightforwardly understood in our standard ΛCDM based view of galaxy formation: voids are expected to be teeming with dwarfs and low surface brightness galaxies. Various astrophysical processes, ranging from gas and radiation feedback processes to environmental properties of dark matter halos, have been suggested [9, 13, 21, 23, 31]. The issue is, however, far from solved and progress will depend largely on new observations that characterize void galaxies and their immediate environment. An additional issue of cosmological interest is whether we can observe the intricate filigree of substructure in voids, expected as the remaining debris of the merging voids and filaments in the hierarchical formation process [7, 10, 28, 32].

Of particular interest in the present context is the manifest environmental influence on the nature of void galaxies. They are found to reside in a more youthful state of star formation. As a population, void galaxies are statistically bluer, have a later morphological type, and have higher specific star formation rates than galaxies in average density environments [11, 22, 26]. Whether void galaxies are intrinsically different or whether their characteristics are simply due to the low mass bias of the galaxy luminosity function in low density regions is still under debate.

An important aspect towards understanding the nature of void galaxies is that of their gas content, about which far less is known than their stellar content. The early survey of 24 IRAS selected IRAS galaxies within the Boötes void by [30] revealed that most of them were gas rich and disk-like, with many gas rich companions. Fresh gas accretion is necessary for galaxies to maintain star formation rates seen today without depleting their observed gas mass in less than a Hubble time [19]. The unique nature of void galaxies provides an ideal chance to distinguish the role of environment in gas accretion and galaxy evolution on an individual basis.

2 The Void Galaxy Survey

The Void Galaxy Survey (VGS) is a multi-wavelength study of ~60 void galaxies, geometrically selected from the SDSS galaxy redshift DR7 survey database (Fig. 1). The project has the intention to study in detail the gas content, star formation history and stellar content, as well as kinematics and dynamics of void galaxies and their companions in a broad sample of void environments. Each of the 60 galaxies has obtained a VGS number, VGS1 until VGS60. Ultimately, we aim at compiling a sample of 50–100 void galaxies.

All galaxies have been selected from the deepest interior regions of identified voids in the SDSS redshift survey on the basis of a unique geometric technique, with no a priori selection on intrinsic magnitude, color or morphology of the void galaxies. The most isolated and emptiest regions in the Local Universe are obtained from the galaxy density and structure maps produced by the DTFE/spine reconstruction technique [1,25,27,34]. From the spatial distribution of the SDSS galaxies, in a volume from $z = 0.003$ to $z = 0.03$, we reconstruct a density field by means of the DTFE procedure, the Delaunay Tessellation Field Estimator. In addition to the computational efficiency of the procedure, the density maps produced by DTFE have the virtue of retaining the anisotropic and hierarchical structures which are so characteristic of the Cosmic Web. The Watershed Void Finder is applied to the DTFE density field for identifying its underdense void basins.

Using the WSRT we have thus far mapped the HI structure of 55 of the 60 galaxies. Of the total VGS sample of 60 void galaxies, the pilot subsample of 15 galaxies has been fully analyzed [17,29]. A necessary sample of comparison galaxies is obtained through simultaneous coverage of regions in front and behind the targeted void galaxies, probing the higher density regions surrounding the targeted void. Note that existing blind HI surveys (ALFALFA, HIPASS) are limited in not resolving the tell-tale HI structures found in the VGS.

In addition to the 5-band photometry and spectroscopy from the SDSS, we obtain deep B- and R-band imaging of all sample galaxies with the La Palma INT telescope and high resolution slit spectroscopy of a subsample of our VGS galaxies. The deep imaging allows us to detect low surface brightness features such as extended, unevolved, stellar disks, tidal streams, the stellar counterparts of several detected HI features (polar rings, tails, etc.) and of the faint HI dwarf companions. Such information is crucial for distinguishing intrinsic formation and evolution scenarios from external processes such as merging and tidal interactions. To probe the old stellar population of the VGS void galaxies, for 10 galaxies we have obtained near-IR JHK WIRC imaging at the 5-m Palomar Hale telescope. In order to assess the distribution of star formation and associated star formation rates, we are obtaining GALEX UV data of 45 galaxies, and have obtained Hα imaging of the complete sample at the MDM telescope.

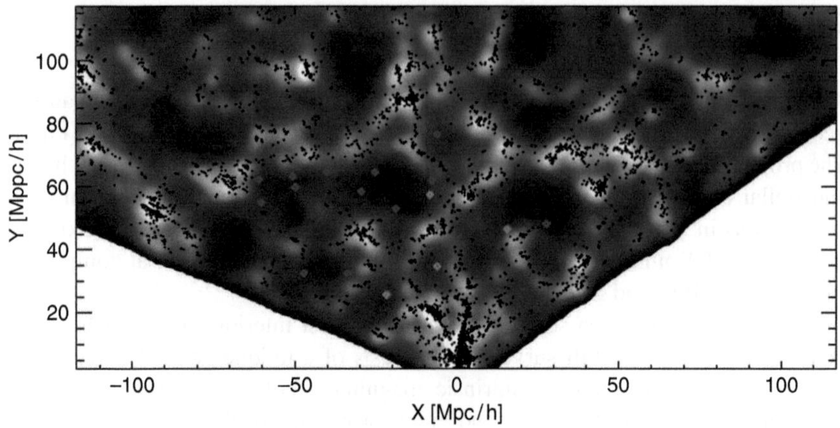

Fig. 1 SDSS density map and identification of voids in the SDSS galaxy redshift survey region from which we selected the galaxies in the Void Galaxy Survey, in a slice $4\,h^{-1}$ Mpc in thickness. The DTFE computed galaxy density map, Gaussian smoothed on a scale of $R_f = 1\,h^{-1}$ Mpc, is represented by the greyscale map. The SDSS galaxies are superimposed as *dark dots*. *Light-grey diamonds*: VGS pilot sample galaxies. *Very dark-grey diamonds*: VGS void galaxies from the full sample. *Dark-grey diamonds*: control sample galaxies (from [17])

3 Results of the VGS Survey: Current State of Affairs

The first results of the Void Galaxy Survey are tantalizing and has revealed a few surprising gas configurations. With a HI mass limit of $\sim 2 \times 10^7\,M_\odot$ and column density limit of $\sim 10^{19}\,\mathrm{cm}^{-2}$, our HI survey provides a significantly improved view of HI in void galaxies compared to past studies [30]. Figure 2 shows one of the most surprising specimen in our survey, the polar ring void galaxy VGS12 (see Sect. 3.3). It is one of the several void galaxies, possibly together with e.g., KK246 and VGS31, that show evidence for cold mode accretion.

The first fully analyzed sample of 15 void galaxies demonstrated the success of our strategy [17]. With 14 detections out of 15, the HI detection rate is very high. We discovered one previously known and five previously unknown companions, while two appear to be interacting. Of these five befriended void galaxies, two are interacting in HI. All HI-detected companions have optical counterparts within the SDSS. Of the nine isolated void galaxies, many exhibit irregularities in the kinematics of their gas disks. Based on their $150\,\mathrm{km\,s}^{-1}$ velocity width, the detected target galaxies have a range of HI masses from $0.35–3.8 \times 10^9\,M_\odot$, Companion galaxies have masses ranging from $0.5–4.5 \times 10^8\,M_\odot$.

While our targeted void galaxies are small, they would not be classified as dwarf galaxies [12]. All have $M_r < -16$ and exhibit small circular velocities of $50–100\,\mathrm{km\,s}^{-1}$. All exhibit signs of rotation, though limiting resolution and lower sensitivity at the disk outskirts means we do not always see a flattening of the

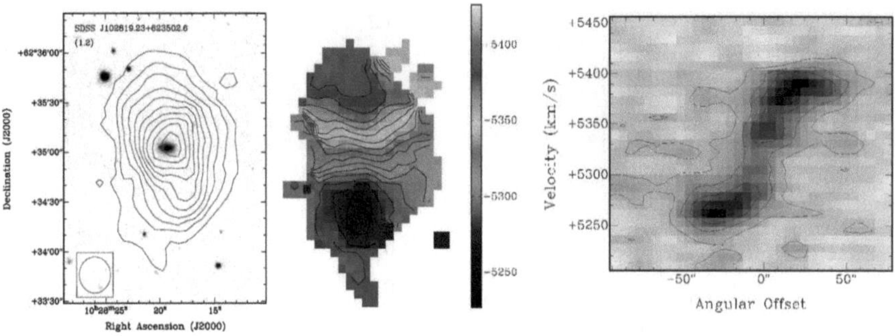

Fig. 2 Targeted void galaxy VGS12, the polar ring void galaxy. *Left*: HI intensity map, superimposed on optical image. Contours are at $5 \times 10^{19}\,cm^{-2}$, plus increments of $10^{20}\,cm^{-2}$. *Centre*: velocity field. *Lines* indicate increments of $8\,\mathrm{km\,s^{-1}}$. *Right*: position-velocity diagram along the kinematic major axis (from [17])

rotation curve. M_{dyn} is typically 10^9–$10^{10}\,M_\odot$. The detected companions are more dwarfish.

The void galaxy population appears to represent the extreme blue and faint tail of an otherwise normal galaxy population. There are a few characteristics which seem to set them apart, mainly concerning their HI gas content and star formation activity.

3.1 Magnitudes and Colors

Despite their affected outer regions, our study finds that in the color-magnitude diagram the target void galaxies nicely reside at the faint end of the blue cloud of galaxies. This can be immediately appreciated from the diagrams in Fig. 3, showing the magnitude and color distribution of our galaxies as a function of density excess/deficit δ. Most of our galaxies find themselves on the blue sequence of SDSS galaxies, towards the bluest edge of these galaxies. We find that our pilot sample galaxies are at the faint end of the galaxy luminosity function (see e.g., also [26])! Because our geometric selection procedure manages to probe the extremely underdense and desolate void interiors, our void galaxy sample is able to probe specifically those low luminosity galaxies which make up the bulk of the void galaxy population and were previously inaccessible (Fig. 3, left). Also apparent is the dominance of blue galaxies at the deepest underdensities (Fig. 3, right).

3.2 Star Formation and Gas Content

A key aspect of the HI observations is that even in these rather desolate and underdense void regions several cases of very irregular HI morphologies were

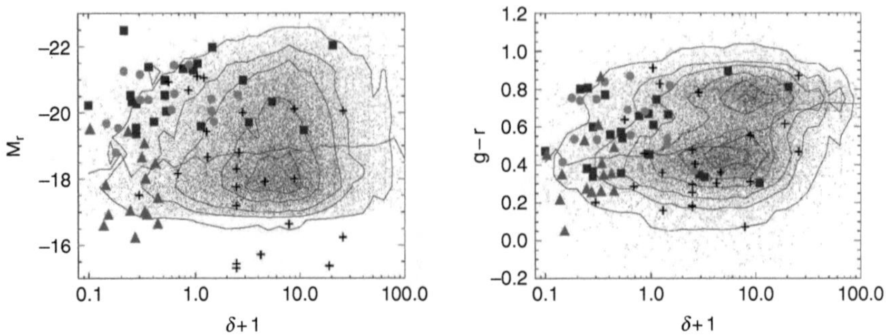

Fig. 3 Magnitudes and color of void galaxies as a function of density excess/deficit δ. *Left*: distribution of r-band absolute magnitudes for our void galaxy sample (*triangles*), the Boötes void galaxy sample of [30] (*squares*) and the CfA void galaxy sample of [11] (*crosses*). These are compared to the general color-magnitude diagram of a volume-limited sample of SDSS galaxies, with $z < 0.02$ and $M_r < -16.9$ (*dots*, from [17]). *Right*: distribution of (g - r) colours as a function of δ

detected, characterized by disturbed HI disks, tails, warps, and cold gas filaments. This suggests that void galaxies are activily building up. A related second key aspect is the high abundance of faint *companions*, non-interacting as well as merging.

One particular aspect in which we find a systematic deviation of our void galaxies, with respect to the norm for similar galaxies, is their size. They have stellar disks that are smaller than average, with the r-band r_{90} radius of our late type void galaxies systematically lower than the median for late type galaxies (Fig. 4, left). However, the result is tentative and might be beset by a hidden selection effect.

Perhaps the most outstanding characteristic of void galaxies is that of their star formation properties. In general, the stellar and star formation properties of our VGS pilot sample are in agreement with the values found in other samples of void galaxies [26]. In this respect, it is relevant that the HI mass of the VGS galaxies appears to be typical in following the global trend of an increasing hydrogen mass M_{HI} as their optical (r-band) luminosity decreases: the smallest galaxies have been less efficient at turning gas into stars. However, when assessing possible trends with density, we find that the specific star formation rate (SFR per stellar mass) of the galaxies displays a suggestive systematic trend. There is a distinct trend for an increase of the star formation rate per HI mass for galaxies in lower density areas (Fig. 4, right).

In all, we find that the outer regions and immediate environment of void galaxies testify of strong recent interactions and star formation activity. This, by itself, is a surprising finding for galaxies populating the most desolate areas of our Universe.

3.3 *Exuberance in the Desert*

An outstanding specimen of our sample is the polar disk galaxy VGS12 [29]. Amongst the most lonely galaxies in the Universe, it has a massive, star-poor HI disk

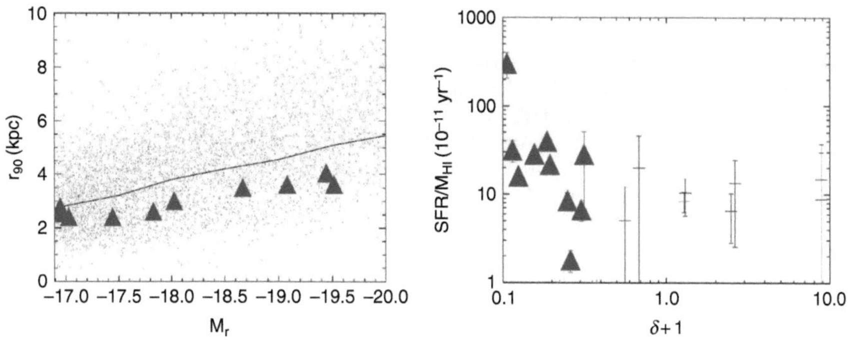

Fig. 4 Two deviant characteristics of VGS void galaxies. *Left*: The r-band r_{90} radii of stellar disks. Our late-type void galaxies (*triangles*) fall systematically below the median (*line*) of a volume limited SDSS sample of late-type galaxies (*dots*). *Right*: star formation rate per hydrogen mass, SFR/M_{HI}, as a function of density of our void galaxies (*triangles*) and our control sample (*crosses*). The void galaxies have a higher star formation rate per hydrogen mass (from [17])

that is perpendicular to the disk of the central void galaxy. No optical counterpart to the HI disk has yet been found, even though the inner optical galaxy is actively forming stars. The galaxy is located within a tenuous wall in between two large roundish voids. The undisrupted appearance of the original stellar disk renders a merger origin unlikely. It suggests slow accretion of cold gas [3, 6, 15], at the crossing point of the outflow from the two voids. Cold accretion as a formation mechanism for polar ring galaxies has been seen to occur in simulations [5, 20].

Another fascinating object is VGS31. It defines a system of three galaxies, stretching out over 57 kpc and possibly connected by a HI bridge. The easternmost object is a Markarian galaxy, marked by prominent stellar streams wrapping around the central galaxy and a separate tidal tail or stream. These might be the remnants of the recent infall of one or two satellite galaxies [2]. The westernmost object is considerably fainter than the other two galaxies. The tails and streams are also visible in the recent deep INT B imaging as well as on the NIR J and K maps, possibly with a slight and unique misalignment. The fact that the gas and all objects involved appear to be stretched along a preferred direction may be suggestive of a system situated within a tenuous filament within the large encompassing void.

References

1. M.A. Aragón-Calvo et al., ApJ **723**, 364 (2010)
2. B. Beygu et al., in preparation (2011)
3. J. Binney, ApJ **215**, 483 (1977)
4. J.R. Bond, L. Kofman, D. Pogosyan, Nature **380**, 603 (1996)
5. C.B. Brook et al. ApJ **689**, 678 (2008)
6. A. Dekel, Y. Birnboim, MNRAS **368**, 2 (2006)

7. J. Dubinski et al., ApJ **410**, 458 (1993)
8. J. Einasto, M. Joeveer, E. Saar, MNRAS **193**, 353 (1980)
9. S.R. Furlanetto, T. Piran, MNRAS **366**, 467 (2006)
10. S. Gottlöber et al., MNRAS 344, **715** (2003)
11. N.A. Grogin, M.J. Geller, AJ **118**, 2561 (1999)
12. P.W. Hodge, Ann. Rev. Astron. Astrophys. **9**, 35 (1971)
13. M. Hoeft et al., MNRAS 371, 401 (2006)
14. F. Hoyle, M.S. Vogeley, ApJ **566**, 641 (2002)
15. D. Kereš et al., MNRAS **363**, 2 (2005)
16. R.P. Kirshner et al., ApJL **248**, L57 (1981)
17. K. Kreckel et al., AJ **141**, 4 (2011)
18. V. de Lapparent, M.J. Geller, J.P. Huchra, ApJL **302**, L1 (1986)
19. R.B. Larson, Nature **236**, 21 (1972)
20. A.V. Macció, B. Moore, J. Stadel, ApJ **636**, 25 (2006)
21. H. Mathis, S.D.M. White, MNRAS **337**, 1193 (2002)
22. S.G. Patiri et al., MNRAS **369**, 335 (2006)
23. P.J.E. Peebles, ApJ **557**, 495 (2001)
24. P.J.E. Peebles, A. Nusser, Nature **465**, 565 (2010)
25. E. Platen, R. van de Weygaert, B.J.T. Jones, MNRAS **380**, 551 (2007)
26. R.R. Rojas et al., ApJ **617**, 50 (2004)
27. W.E. Schaap, R. van de Weygaert, A&A **363**, L29 (2000)
28. R.K. Sheth, R. van de Weygaert, MNRAS **350**, 517 (2004)
29. K. Stanonik et al., ApJL **696**, L6 (2009)
30. A. Szomoru et al., AJ **111**, 2150 (1996)
31. J.L. Tinker, C. Conroy, ApJ **691**, 633 (2009)
32. R. van de Weygaert, E. van Kampen, MNRAS **263**, 481 (1993)
33. R. van de Weygaert, E. Platen, Modern Phys. Lett. A. (2009, in press). arXiv:0912.2997
34. R. van de Weygaert, E. Schaap, in *LNP 665*, ed. by V.J. Martínez, E. Saar, E. Martínez-Gonzlez, M.-J. Pons-Bordería. Data Analysis in Cosmology (Springer, Heidelberg, 2009), pp. 291–413

Metallicities of Galaxies in the Lynx-Cancer Void

A. Kniazev, S. Pustilnik, A. Tepliakova, and A. Burenkov

Abstract Does the void environment have a sizable effect on the evolution of dwarf galaxies? If yes, the best probes should be the most fragile least massive dwarfs. We compiled a sample of about one hundred dwarfs with M_B in the range −12 to −18 mag, falling within the nearby Lynx-Cancer void. The goal is to study their evolutionary parameters – gas metallicity and gas mass-fraction, and to address the epoch of the first substantial episode of Star Formation. Here we present and discuss the results of O/H measurements in 38 void galaxies, among which several of the most metal-poor galaxies are found with an oxygen abundance of $12 + \log(O/H) = 7.12$–7.3 dex.

1 Objectives

In the framework of Cold Dark Matter models, dwarf galaxies in voids could form later and evolve more slowly than their counterparts in a more typical environment. However, quantitative predictions are uncertain. Data on the evolution of the void population are rather scarce and indirect. We recently studied a nearby void in Lynx-Cancer (Pustilnik and Tepliakova, 2010, MNRAS, submitted) with $D_{cent} \sim 18$ Mpc and the size ~ 16 Mpc. About 100 dwarfs fall in this void. The goal of the ongoing project is to obtain the evolutionary parameters of the void sample galaxies. Here we present the intermediate results of O/H determination, based mainly on the SAO 6-m telescope observations.

A. Kniazev (✉)
SAAO, P.O. Box 9, 7935 Observatory, Cape Town, South Africa
e-mail: akniazev@saao.ac.za

S. Pustilnik · A. Tepliakova · A. Burenkov
SAO, Nizhnij Arkhyz, Karachai-Circassia, 369167, Russia
e-mail: sap@sao.ru; arina@sao.ru; ban@sao.ru

I. Ferreras and A. Pasquali (eds.) *Environment and the Formation of Galaxies: 30 years later*, Astrophysics and Space Science Proceedings,
DOI 10.1007/978-3-642-20285-8_4,

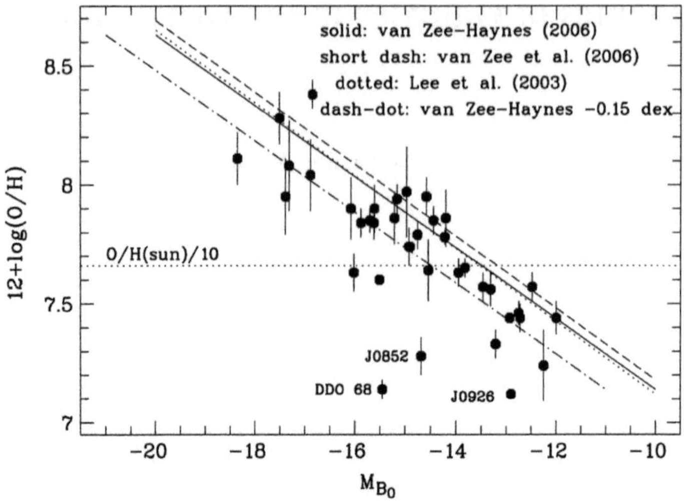

Fig. 1 Metallicity-Luminosity relation for 38 Lynx-Cancer void galaxies. *Solid*, *short-dashed* and *dotted lines* show respectively the known L–Z relations for isolated late-type galaxies [6], nearby dIs [7] and nearby dI and I galaxies from [3]. The *dash-dotted line* is 0.15 dex lower than the *solid* one (for which the fit rms = 0.15 dex). Many L–C void galaxies (especially those with $M_B > -16.0$) fall below the *dash-dotted line*

2 Results

Currently we have oxygen abundances for 38 Lynx-Cancer void galaxies, which are shown as the diagram O/H versus M_B in Fig. 1. The majority of galaxies have O/H derived via classic T_e-method with programs described in [2]. About 1/3 of dwarfs with faint or undetected [OIII] λ4,363 lines have O/H derived via the semi-empirical method of [1]. Three lines, close to each other, show fits to similar empirical relations for local I/dI galaxies and isolated late-type galaxies from [3, 6] and [7]. The dash-dotted line is just shifted down by 0.15 dex relative to that by [6]. About 1/3 of void galaxies have O/H below this line. The effect looks more prominent for $M_B > -16$. The latter could be the first indication for the slower chemical evolution of the least massive galaxies in voids.

3 Unusual Dwarfs in Lynx-Cancer Void

In the course of this void galaxy sample study, an unusual concentration of the most metal-poor dwarfs is found. Namely, at least five galaxies with $12+\log(O/H) \leq 7.3$ fall in this void: J0926+3343 (7.12) [5], DDO 68 (7.14) [1, 4], J0737+4724 (7.24), J0852+1350 (7.28), J0812+4836 (7.28) [1]. In contrast to the majority of the void galaxies, the former three have SDSS colours in their outer parts, indicating no traces of stars with ages older than 1–3 Gyr. Two more Lynx-Cancer void LSBDs, SAO 0822+3545 and SDSS J0723+3622, show no traces of an older stellar population. The void sample shows the sizable overabundance of the most metal-poor objects compared to the Local Volume sample. This fact also suggests that void dwarfs may experience a slower chemical evolution.

References

1. Y.I. Izotov, T.X. Thuan ApJ **665**, 1115 (2007)
2. A. Kniazev et al., MNRAS **388**, 1667 (2008)
3. H. Lee et al., AJ **125**, 146 (2003)
4. S.A. Pustilnik et al., A&A **443**, 91 (2005)
5. S.A. Pustilnik et al., MNRAS **401**, 333 (2010)
6. L. van Zee, M. Haynes ApJ **636**, 214 (2006)
7. L. van Zee et al., ApJ **637**, 269 (2006)

The Dependence of Low Redshift Galaxy Properties on Environment

S.M. Weinmann, F.C. van den Bosch, and A. Pasquali

Abstract We review recent results on the dependence of various galaxy properties on environment at low redshift. As environmental indicators, we use group mass, group-centric radius, and the distinction between centrals and satellites; examined galaxy properties include star formation rate, colour, AGN fraction, age, metallicity and concentration. In general, satellite galaxies diverge more markedly from their central counterparts if they reside in more massive haloes. We show that these results are consistent with starvation being the main environmental effect, if one takes into account that satellites that reside in more massive haloes and at smaller halo-centric radii on average have been accreted a longer time ago. Nevertheless, environmental effects are not fully understood yet. In particular, it is puzzling that the impact of environment on a galaxy seems independent of its stellar mass. This may indicate that the stripping of the extended gas reservoir of satellite galaxies predominantly occurs via tidal forces rather than ram-pressure.

1 Introduction

It has long been known that galaxies living in dense regions tend to be redder, less active in their star formation and of earlier type [3, 8, 31]. More recently, large galaxy surveys like the SDSS and reliable stellar mass estimates have indicated that

S.M. Weinmann (✉)
Leiden Observatory, Leiden University, P.O. Box 9513, 2300 RA Leiden, The Netherlands
e-mail: weinmann@strw.leidenuniv.nl

F.C. van den Bosch
Astronomy Department, Yale University, P.O. Box 208101, New Haven, CT 06520-8101, USA
e-mail: frank.vandenbosch@yale.edu

A. Pasquali
Astronomisches Rechen-Institut, Zentrum für Astronomie der Universität Heidelberg, Mönchhofstrasse 12-14, 69120 Heidelberg, Germany
e-mail: pasquali@ari.uni-heidelberg.de

I. Ferreras and A. Pasquali (eds.) *Environment and the Formation of Galaxies: 30 years later*, Astrophysics and Space Science Proceedings,
DOI 10.1007/978-3-642-20285-8_5,

the main parameter governing galaxy properties is however not environment, but stellar mass [20]. It is therefore crucial that environmental effects are studied at fixed stellar mass.

Traditionally, environment has been described in terms of galaxy density (for example out to the n-th nearest neighbour), or in terms of the field versus cluster distinction. Group and cluster finding algorithms applied to the SDSS and other surveys have now made it possible to quantify environment in a way that is physically better motivated, expressing it in terms related to the expected underlying dark matter structure [4, 17, 39, 46]. To first order, one can quantify environment simply by discriminating between central and satellite galaxies [36], which acknowledges the fact that those two kinds of galaxies have different average dark matter accretion histories. Finer distinctions between different subpopulation of satellites can then be made according to their host halo mass and group- or cluster-centric distance.

Taking out stellar mass dependencies, and describing environment with a physically better motivated language makes it easier to interpret the observed environmental dependencies. Many different processes have been suggested to drive these environmental dependencies, including ram-pressure stripping of the cold gas [14], removal of the extended (hot) gas reservoir of galaxies by tidal or ram-pressure stripping ("starvation," [3,23]), harassment by high-speed tidal encounters [25, 29], or a faster mass growth at early times for galaxies that end up in more massive haloes at late times [30]. Much of the recent work on this topic converges towards "starvation" being the main driver of environmental effects [38, 40, 41]. This is an outcome foreseen by early semi-analytical models which included no other environmental effect except the instantaneous removal of the satellite's hot gas reservoir at accretion [19]. However, it has also been shown that this simple recipe likely makes starvation in semi-analytical models overefficient [26, 42], and there are still open questions on how and on which timescales this effect exactly operates.

In this review, we summarize recent results on how various galaxy properties depend on environment at fixed stellar mass. We then outline potential implications of these results on our understanding of galaxy evolution as a function of environments.

2 SDSS Group and Cluster Catalogues

Most of the results described in what follows are based on the Yang et al. [46] group catalogue, which makes use of the SDSS-DR4 [1] and the New York Value-Added Galaxy Catalogue [5]. This group catalogue is constructed using the halo-based group finder of [45], which uses an iterative scheme and priors on the redshift-space structure of dark matter haloes to partition galaxies over groups. Halo masses are estimated from a ranking in total characteristic luminosity or total characteristic stellar mass, with both methods giving very similar results [46]. Details on the galaxy group sample used in most of the studies mentioned below can be found in [35]. Some results come from the cluster catalogue by von der Linden et al. [39], which is based on the SDSS DR4 catalogue and the C4 cluster catalogue [27]. In what follows, a "central" galaxy is defined as the most massive galaxy in its group. All other galaxies which reside in the same group are labelled "satellites."

3 Results

3.1 The Dependence of Galaxy Properties on the Satellite-Central Dichotomy

To first order, we can quantify environmental dependence simply by comparing satellite and central galaxies of the same stellar mass. Several key differences between those two types of galaxies have been found. The most basic difference is that satellite galaxies are redder [35] and have lower specific star formation rates [22] than central galaxies of the same stellar mass. In addition, satellite galaxies are less likely to reveal optical or radio AGN activity, and their AGN activity is weaker than in their central counterparts [32]. In terms of morphology, it is found that satellite galaxies have surface brightness profiles that are, on average, slightly more concentrated than those of central galaxies [16, 35, 43]. As shown by [43], this most likely does not reflect a true difference in the mass distribution, but rather can be explained by fading of the stellar disk due to star formation quenching in the satellites. This is consistent with the finding that satellites and centrals reveal no structural differences if they are matched in both stellar mass and colour [16, 35]. As can be seen in Fig. 1, the fraction of galaxies with a concentration > 3 is the same for satellites and centrals at fixed stellar mass. This suggests that the most

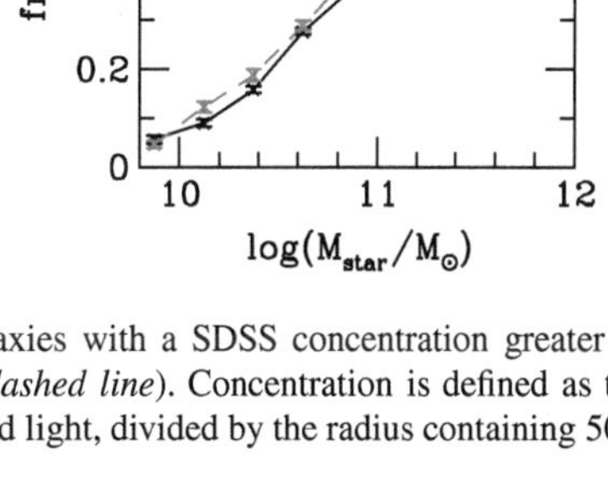

Fig. 1 The fraction of galaxies with a SDSS concentration greater than 3 for central galaxies (*solid line*) and satellites (*dashed line*). Concentration is defined as the ratio between the radius containing 90% of the r-band light, divided by the radius containing 50% of the r-band light. More details are given in [43]

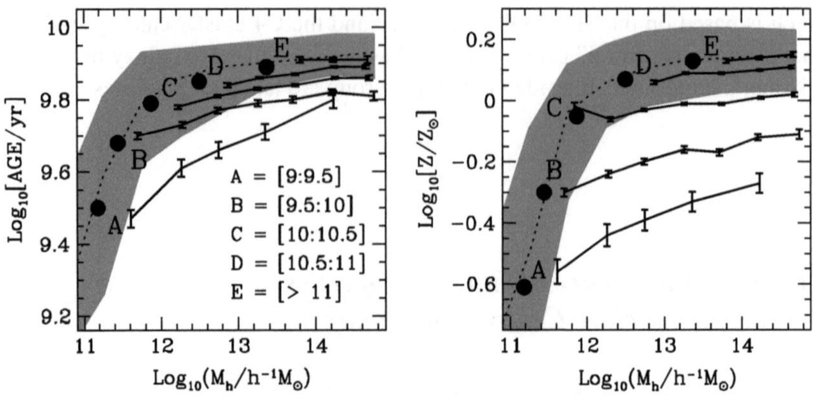

Fig. 2 The mean stellar mass-weighted age (*left hand panel*) and metallicity (*right hand panel*) for central galaxies (*large filled circles*) and for satellite galaxies (*lines with errorbars*) for different stellar mass bins A–E, as indicated. Results for satellites are shown as a function of halo mass. It can be seen that low mass centrals have lower metallicities than satellites, and that the metallicity for low mass satellites increases with increasing host halo mass. More details are given in [33]

highly concentrated galaxies (which typically have elliptical morphologies) are not produced by environmental effects.

3.2 *The Dependence of Satellite Properties on Host Halo Mass*

Group catalogues make it possible to study the properties of satellite galaxies as a function of the mass of their host halo. While the average colours of satellites depend only very weakly on host halo mass [36], there seems to be a significant increase of the fraction of red or passive satellites with increasing halo mass [22, 38]. Interestingly, satellites in more massive haloes are older and more metal rich than their counterparts in lower mass haloes [33]. This is shown in Fig. 2, which plots the stellar mass-weighted ages and metallicities of centrals and satellites as a function of host halo mass. The halo mass dependence (and hence the difference between centrals and satellites) is most pronounced at relatively low stellar masses, and disappears at the high mass end [33].

While the older ages of satellites are likely directly related to their larger passive fraction, their higher metallicity is less straightforward to explain. It could indicate that the satellites are the descendants of centrals with higher stellar mass, and thus a higher metallicity, which underwent tidal stripping of their stellar material [33]. However, the amount of stellar mass stripping required is fairly substantial, which is difficult to reconcile with the fact that satellites seem to have the same concentrations like centrals of the same stellar mass (see for example [27]). Another explanation could be that satellites form a significant amount of stars after infall (but see [2]). It is plausible that these would have a higher metallicity than stars formed in a comparable central galaxy due to the lack of new infall of low metallicity ('primordial') gas. However, a significant boost in the stellar mass-weighted metallicity requires that satellites form a relatively large fraction of their stars after accretion, which may proof difficult to reconcile with their high passive fractions. Clearly the origin of the relatively high metallicities in low mass satellites needs to be investigated in more detail, and might give interesting new insights into the processing and recycling of gas in both central and satellite galaxies.

Finally, although there are indications from HOD modelling that there is a relation between the fraction of satellites with AGN activity and host halo mass [28], no such relation has been found using our group catalogue [32].

3.3 *The Dependence of Satellite Properties on Cluster-Centric Radius*

Dependencies of galaxy properties on group- or cluster-centric distance are most clearly visible in massive clusters where statistics are best and the center of the cluster is easier to define than in poor groups. Figure 3 shows the fraction of passive satellites in haloes with $M > 10^{14}\,h^{-1}\,M_{\odot}$ as a function of halo-centric radius in the sample used by [44]. Clearly, in these massive haloes the passive fraction of satellites at a given stellar mass increases towards the center [4,40,44]. Interestingly, the fraction of galaxies showing signs of *fast recent* truncation of star formation is found to be virtually independent of cluster-centric radius [40], which seems to suggest that *fast* truncation may be unrelated to environmental effects. Finally, the

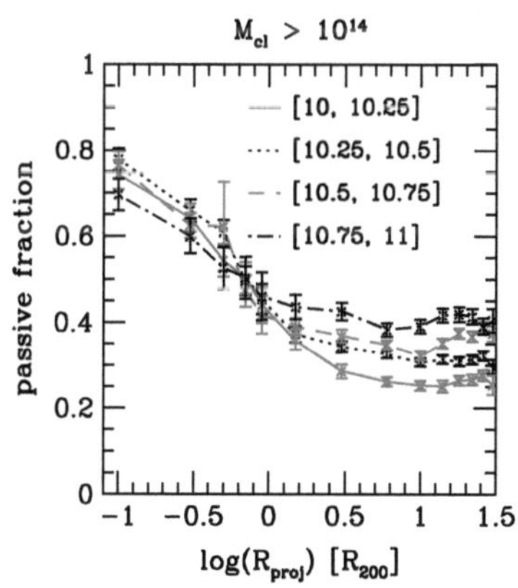

Fig. 3 The fraction of passive satellites in clusters with halo masses $M > 10^{14}\,h^{-1}\,M_{\odot}$, obtained from [39], as a function of projected cluster-centric radius. Results are shown for four different stellar mass bins, as indicated [values in brackets refer to $\log(M_{star}/h^{-2}\,M_{\odot})$]. Here "passive" is defined as having a SSFR $< 10^{-11}yr^{-1}$. Note the clear decrease with increasing cluster-centric radius. For more details see [44]

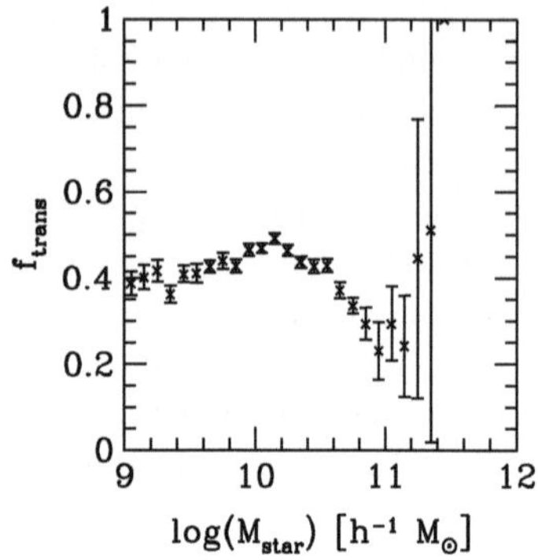

Fig. 4 The transition fraction f_{trans}, which expresses the blue-to-red transition fraction of satellites that were still blue at the time of accretion (see [1]), as a function of stellar mass. A galaxy is defined as red if ${}^{0.1}(g-i) > 0.76 + 0.15\cdot[\log(M_{\mathrm{star}}/\mathrm{h}^{-1}\,M_{\odot}) - 10.0]$. See [35] for details

fraction of galaxies hosting a powerful optical AGN has been found to decrease towards the cluster center at fixed stellar mass [40].

3.4 *The Puzzling Independence of Environmental Effects on Stellar Mass*

An interesting question that has not received much attention yet is how the impact of environment depends on the stellar mass of the galaxy. Naively, one would expect that stripping of the hot or cold gas in a galaxy by any kind of effect is easier for a shallower potential well, and therefore that low mass galaxies are more vulnerable to environmental effects. To investigate the dependence of the strength of environmental effects on stellar mass, we can introduce a quantity f_{trans}, which gives us an estimate on the fraction of galaxies which were blue at the time of infall and have by now become red due to environmental effects [35]:

$$f_{\mathrm{trans}} = \frac{f_{\mathrm{sat,red}} - f_{\mathrm{cen,red}}}{f_{\mathrm{cen,blue}}}, \tag{1}$$

with $f_{\mathrm{sat,red}}$ and $f_{\mathrm{cen,red}}$ the red fraction of satellites and centrals respectively, and $f_{\mathrm{cen,blue}}$ the blue fraction of centrals. Note that we assume here that the blue fraction of centrals today corresponds to the blue fraction of the satellites at the time of infall, which should be roughly correct, since most satellites fall in relatively late [35]. As shown in Fig. 4, f_{trans} is remarkably constant at around 40% from $10^9 M_{\odot}$ up to $10^{11} M_{\odot}$. A similar result was obtained by [34]. These findings imply that the probability for a galaxy to become red due to its environment is nearly independent of its stellar mass, which is challenging to understand from a theoretical perspective.

4 Discussion

The fact that satellites and centrals are different is not surprising given that the former are believed to reside in dark matter subhaloes orbiting within a larger dark matter halo (the "host" halo), while centrals are expected to reside at rest at the center of a host halo. Whereas host haloes continue to grow in mass via accretion, subhaloes lose mass (to their host) due to tidal stripping. This implies that subhaloes, and their associated satellites, also lack the accretion of intergalactic gas associated with halo growth. However, this effect by itself does not necessarily mean that centrals and satellites are different. After all, if satellite galaxies can hang on to their extended haloes of hot gas which they are predicted to have at infall, they may remain "active" for a relatively long period of time [44]. An additional environmental effect therefore seems required to explain the observed dependencies. Based on all results discussed above, this effect has to have the following properties:

- It is likely to result in a depletion of gas, which directly causes higher passive fractions, lower AGN fractions, and higher average stellar-mass weighted ages in satellites compared to centrals of the same stellar mass.
- It occurs for satellite galaxies in haloes spanning a large range in mass (not only in clusters), but becomes weaker towards less massive haloes, and towards the outskirts of clusters [4, 22, 40, 44].
- It quenches star formation in galaxies on a rather long timescale of the order of 2–3 Gyr [18, 38, 40, 43].
- There are no indications that it results in an accompanying structural transformation [16, 37, 43].
- It is similarly strong for galaxies with different stellar masses [34, 35].

Of all the environmental effects that have been suggested, starvation matches this description best. This effect should take place in all kinds of groups, not only in massive clusters. Its effect on the star formation rate is slow, and it does not lead to any morphological changes in the galaxy. However, how this effect operates in detail is still under debate. Also, it is not a priori clear why it should be similarly strong for galaxies with different stellar masses.

Semi-analytical models that include instantaneous and complete removal of the extended gas reservoir of satellites upon accretion result in a passive or red fraction of satellites that is much too high [42]. It has therefore become clear that starvation has to be a more gradual process, which is taken into account in some of the newest semi-analytical models [10, 15, 44]. However, there are indications that modelling ram-pressure stripping of the hot halo according to standard prescriptions leads to a too high fraction of low mass, passive galaxies [44] and also produces too many satellite galaxies with intermediate colours [3, 44]. This could indicate that starvation (i.e. the depletion of the hot gas reservoir) mainly occurs by tidal stripping, and not by ram-pressure stripping [44]. Another argument for this is that tidal stripping seems to be less strongly stellar mass dependent than ram-pressure

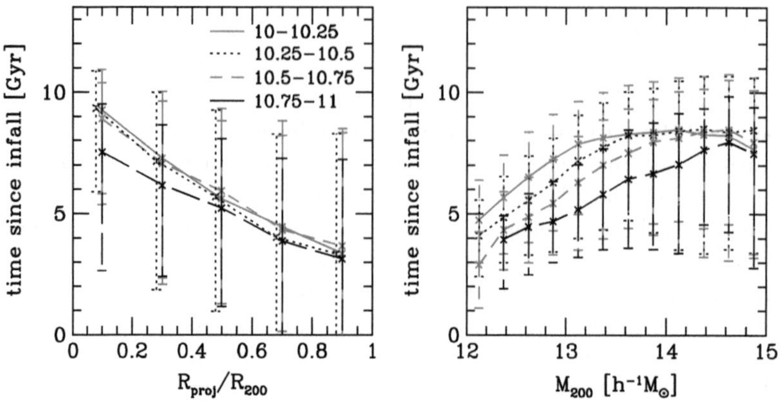

Fig. 5 The median time since galaxies switched from the central to the satellite status for the last time in a semi-analytical model [7], for four different stellar mass bins, as indicated. In the *left hand panel*, we use a sample of clusters selected in a similar way as in observations, as described in detail in [44], and plot the median time since infall as a function of projected cluster-centric distance. Cluster masses are estimated according to velocity dispersion, as explained in [44]. We only use clusters with "observed" masses of 10^{14}–$10^{15}\,h^{-1}M_\odot$. In the *right hand panel*, we plot the median time since infall as a function of group mass, for groups as selected directly in the Millennium simulation. *Errorbars* in both panels denote the range where 68% of the measured values lie

stripping [44], which likely helps in reproducing the stellar-mass independence of environmental effects discussed above. It also does not seem entirely unrealistic that ram-pressure stripping could be overestimated in standard semi-analytical models, as these tend to overestimate the hot gas content of groups [6, 44].

The observation that galaxies in cluster centers and in more massive groups have a higher passive fraction than their counterparts in the outskirts and in less massive groups can have two different implications. Either environmental effects are simply strongest in group centers and in massive clusters, or galaxies residing at smaller halo-centric distances and/or in more massive haloes have on average been satellites for a longer period so that environmental effects have had more time to operate. Indeed, for the semi-analytical model of [7], we find that satellites in more massive systems, and closer to the projected cluster-center, have been satellites for longer period, which is shown in Fig. 5. Neither dependency is surprising. Galaxies that end up in more massive haloes are "born" in denser environments, and therefore tend to become satellites at earlier times. Also, it takes time for newly infalling galaxies to sink to the cluster center by dynamical friction, which explains the radial dependence (see also [12, 13]). Hence, based on Fig. 5, we conclude that the larger fractions of passive satellites in cluster centers and in massive clusters most likely reflect trends in the time of infall, and do not require that environmental effects are stronger in more massive haloes and/or at smaller halo-centric radii.

5 Outlook

Models of environmental effects start to be able to reproduce the relations between galaxy properties and environment for low redshift galaxies with masses above $10^9\,M_\odot$ [44], indicating that our understanding of environmental effects is improving. Of course, these models should be further tested and refined by using more detailed observational data at low redshift. For example, the gas content of galaxies in different environments can hold important additional clues on the interplay between accretion, SN feedback and environmental effects [21].

Another important test for these models is whether or not they can capture the redshift evolution of environmental dependencies. However, at high redshift there are still several important discrepancies between current semi-analytic model predictions and galaxy properties in general [9, 11, 15] which may need to be addressed first.

Finally, it is important to probe environmental effects at masses lower than discussed here. Although the efficiency of environmental effects seems to be nearly independent of stellar mass in the mass range discussed here, this might well change at even lower stellar masses. To explain the observed population of dwarf elliptical galaxies and their different subclasses as found for example in the Virgo cluster [24], processes additional to starvation, like harassment, or a different formation channel at early times, might be required.

Acknowledgements We thank all our collaborators on this topic, in particular Houjun Mo, Xiaohu Yang, Guinevere Kauffmann, Anja von der Linden, Gabriella De Lucia, Anna Gallazzi, Fabio Fontanot, Yicheng Guo, Daniel McIntosh, and Xi Kang.

References

1. J.K. Adelman-McCarthy et al., ApJ **162**, 38 (2006)
2. M.L. Balogh et al., ApJ **488**, L75 (1997)
3. M.L. Balogh et al., MNRAS **398**, 754 (2009)
4. S.P. Bamford et al., MNRAS **393**, 1324 (2009)
5. M.R. Blanton et al., AJ **129**, 2562 (2005)
6. R.G. Bower, I.G. McCarthy, A.J. Benson, MNRAS **390**, 1399 (2008)
7. G. De Lucia, J. Blaizot, MNRAS **375**, 2 (2007)
8. A. Dressler, ApJ **236**, 351 (1980)
9. I. Ferreras et al., MNRAS **396**, 1573 (2009)
10. A.S. Font et al., MNRAS **389**, 1619 (2008)
11. F. Fontanot et al., MNRAS **397**, 1776 (2009)
12. J. Gan et al., MNRAS **408**, 2201 (2010)
13. L. Gao et al., MNRAS **355**, 819 (2004)
14. J.E. Gunn, J.R. Gott III, ApJ **176**, 1 (1972)
15. Q. Guo et al., MNRAS **413**, 101 (2011). arXiv:1006.0106
16. Y. Guo et al., MNRAS **398**, 1129 (2009)
17. S.M. Hansen et al., ApJ **699**, 1333 (2009)
18. X. Kang, F.C. van den Bosch, ApJ **676**, 101 (2008)

19. G. Kauffmann, S.D.M. White, B. Guideroni, MNRAS **264**, 201 (1993)
20. G. Kauffmann et al., MNRAS **341**, 33 (2003)
21. G. Kauffmann, L. Cheng, T.M. Heckman, MNRAS **409**, 491 (2010). arXiv:1005.1825
22. T. Kimm et al., MNRAS **394**, 1131 (2009)
23. R.B. Larson, B.M. Tinsley, C.N. Caldwell, ApJ **237**, 692 (1980)
24. T. Lisker et al., ApJ **660**, 1186 (2007)
25. C. Mastropietro et al., MNRAS **364**, 607 (2005)
26. I. McCarthy et al., MNRAS **383**, 593 (2008)
27. C.J. Miller et al., AJ **130**, 968 (2005)
28. T. Miyaji et al. ApJ **726**, 83 (2011). arXiv:1010.5498
29. B. Moore et al., Nature **376**, 613 (1996)
30. E. Neistein et al. MNRAS (2011, in press). arXiv:1011.2492
31. A. Oemler, ApJ **194**, 1 (1974)
32. A. Pasquali et al., MNRAS **394**, 38 (2009)
33. A. Pasquali et al., MNRAS **407**, 937 (2010)
34. Y. Peng et al., ApJ 721, 193 (2010)
35. F.C. van den Bosch et al., MNRAS **387**, 79 (2008a)
36. F.C. van den Bosch et al., (2008b, preprint). arXiv:0805.0002
37. A. van der Wel, ApJ **675**, 13 (2008)
38. A. van der Wel et al., ApJ **714**, 1779 (2010)
39. A. von der Linden, P.N. Best, G. Kauffmann, S.D.M. White, MNRAS **379**, 894 (2007)
40. A. von der Linden et al., MNRAS **404**, 1231 (2010)
41. S.M. Weinmann et al., MNRAS **366**, 2 (2006a)
42. S.M. Weinmann et al., MNRAS **372**, 1161 (2006b)
43. S.M. Weinmann et al., MNRAS **394**, 1213 (2009)
44. S.M. Weinmann et al., MNRAS **406**, 2249 (2010)
45. X. Yang et al., MNRAS **356**, 1293 (2005)
46. X. Yang et al., ApJ **671**, 153 (2007)

The VIMOS VLT Deep Survey: A Homogeneous Galaxy Group Catalogue Upto $z \sim 1$

O. Cucciati, C. Marinoni, A. Iovino, and VVDS Collaboration

Abstract We present a homogeneous and complete catalogue of optical galaxy groups identified in the VIMOS-VLT spectroscopic Deep Survey (VVDS). We use mock catalogues extracted from the Millennium simulation to study the potential systematics that might affect the overall distribution of the identified systems. We train on these mock catalogues the adopted group-finding technique (the Voronoi-Delaunay Method, VDM), to recover in an unbiased way the redshift and velocity dispersion distributions of groups and maximise the level of completeness (C) and purity (P) of the group catalogue. We identify 318(/144) VVDS groups with at least 2(/3) members within $0.2 \leq z \leq 1.0$, globally with C = 60% and P = 50%. We use the group sample to study the redshift evolution of the fraction f_b of blue galaxies within $0.2 \leq z \leq 1$ in both groups and in the total VVDS galaxy sample.

1 This Work

We used the data collected for the VVDS-0226-04 deep field ("VVDS-02h" field, [6]) to produce a homogeneous and complete optically-selected galaxy group catalogue, with reliable properties like redshift and line of sight velocity

O. Cucciati (✉)
Laboratoire d'Astrophysique de Marseille, UMR 6110 CNRS-Université de Provence,
38 rue Frederic Joliot-Curie, 13388 Marseille Cedex 13, France
e-mail: olga.cucciati@oamp.fr

C. Marinoni
Centre de Physique Théorique, UMR 6207 CNRS-Université de Provence, 13288, Marseille, France
e-mail: Christian.Marinoni@cpt.univ-mrs.fr

A. Iovino
INAF-OABrera, Via Brera 28, 20121 Milano, Italy
e-mail: angela.iovino@brera.inaf.it

I. Ferreras and A. Pasquali (eds.) *Environment and the Formation of Galaxies: 30 years later*, Astrophysics and Space Science Proceedings,
DOI 10.1007/978-3-642-20285-8_6, © Springer-Verlag Berlin Heidelberg 2011

dispersions (σ_{los}). The aim is to study galaxy properties in very dense regions, on scales at which the physical processes driven by galaxy interactions are believed to play a major role. This work is extensively described in [1]. We refer the reader to [3–5] for similar works based on other data sets.

We used galaxy mock catalogues that mimic the VVDS observational strategy, and that were extracted from the Millennium simulations [8], coupled to a semi-analytic model for galaxy formation [2]. With these mocks, we tested how well the virial l.o.s. velocity dispersion (σ_{vir}) of the halo mass particles can be estimated using sparsely sampled galaxy velocities. We found that in this way, given the characteristics of our survey (flux limit, sampling rate...) we are able to recover a sensible value of σ_{vir} for $\sigma_{vir} \geq 350\,\mathrm{km\,s^{-1}}$. Finally, we used these mock catalogues to train our group-finding algorithm, based on the Voronoi-Delaunay Method (VDM, [7]).

Applying the optimised algorithm to the observed VVDS sample, we obtained a catalogue of 318(/144) groups of galaxies with at least 2(/3) members within $0.2 \leq z \leq 1.0$. The group catalogue is characterised by an overall completeness of ~60% and a purity of ~50%. Nearly 19% of the total population of galaxies live in these systems. We verified that the number density distribution as a function of both redshift and velocity dispersion of the VVDS groups scales in qualitative agreement with the analogous statistics recovered from the mocks.

Finally, we studied the fraction f_b of blue galaxies ($U - B \leq 1$) in the range $0.2 \leq z \leq 1$ (Fig. 1). We used a luminosity-limited subsample of galaxies extracted from our data, complete up to $z = 1$. We found that f_b is significantly lower in

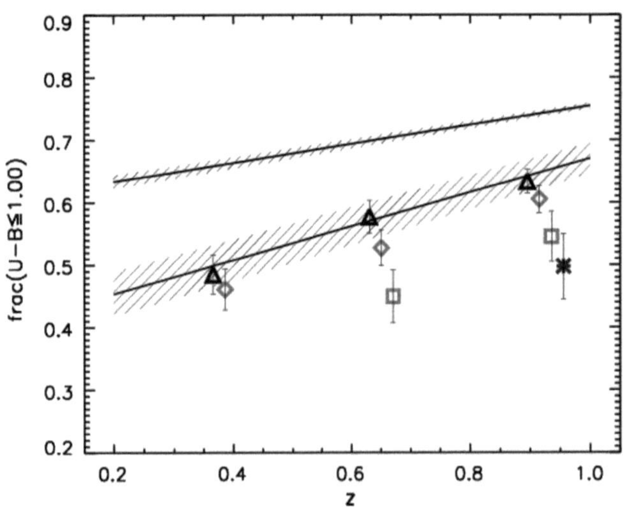

Fig. 1 Fraction f_b of blue galaxies ($U - B \leq 1$) as a function of redshift for group galaxies (*triangles*). The linear fit to these points is over-plotted as a *solid line*, while the *upper black line* is the linear fit for f_b computed within the "total" sample. Other symbols are for group galaxies in groups with increasing richness (from *triangles* to *stars*)

groups than in the global galaxy population. Moreover, f_b increases as a function of redshift irrespective of the environment, with marginal evidence of a higher growth rate in groups. We also found that, at any redshift explored, f_b decreases in systems with increasing richness. Further analyses are left to future work. We only note that the cross-correlation studies of our optically-selected catalogue with samples extracted in the same field with independent techniques will help us to gain insights into cluster selection biases and into the physics at work within these extreme environments.

References

1. O. Cucciati et al., A&A **520**, 42 (2010)
2. G. De Lucia, J. Blaizot, MNRAS **375**, 2 (2007)
3. B.F. Gerke et al., MNRAS **376**, 1425 (2007)
4. A. Iovino et al., A&A **509**, A40+ (2010)
5. C. Knobel et al., ApJ **697**, 1842 (2009)
6. O. Le Fèvre et al., A&A **439**, 845 (2005)
7. C. Marinoni et al., ApJ **580**, 122 (2002)
8. V. Springel et al., Nature **435**, 629 (2005)

The Fossil Candidate RX J1548.9+0851

P. Eigenthaler and W.W. Zeilinger

Abstract Numerical simulations as well as optical and X-ray observations have shown that poor groups of galaxies can evolve to what is called a fossil group. Dynamical friction as the driving process leads to the coalescence of individual galaxies in ordinary poor groups, leaving behind nothing more than a central, massive elliptical galaxy supposed to contain the merger history of the whole group. Due to longer merging timescales for less-massive galaxies, a surrounding faint-galaxy population and an extended X-ray halo similar to that found in ordinary poor groups, is expected. Multi-object spectroscopy with VIMOS has been performed to study the faint galaxy population of the fossil candidate RX J1548.9+0851.

1 The RX J1548.9+0851 System

See [1] for a detailed description on the observational definition of a fossil. Recent simulations suggest that fossils formed at early epochs and have assembled half of their final dark matter mass already at $z \simeq 1$ whereas non-fossils form later on average. This early assembly leaves sufficient time for L^* galaxies to merge into the central one by dynamical friction, resulting in a large magnitude gap at $z = 0$.

We have used VIMOS multi-object spectroscopy to study the optical properties of the fossil candidate RX J1548.9+0851 from the sample of [3]. If the system has indeed formed as proposed by theory, then only the signs of minor mergers are expected to be observed today. Indeed, the RX J1548.9+0851 central elliptical reveals symmetric shells along the galaxy major axis indicative of a recent minor merger (see Fig. 1), supporting the proposed scenario. Forty-one group members

P. Eigenthaler (✉) · W.W. Zeilinger
Institut für Astronomie, Universität Wien, Türkenschanzstraße 17, 1180 Vienna, Austria
e-mail: paul.eigenthaler@univie.ac.at

I. Ferreras and A. Pasquali (eds.) *Environment and the Formation of Galaxies: 30 years later*, Astrophysics and Space Science Proceedings,
DOI 10.1007/978-3-642-20285-8_7,

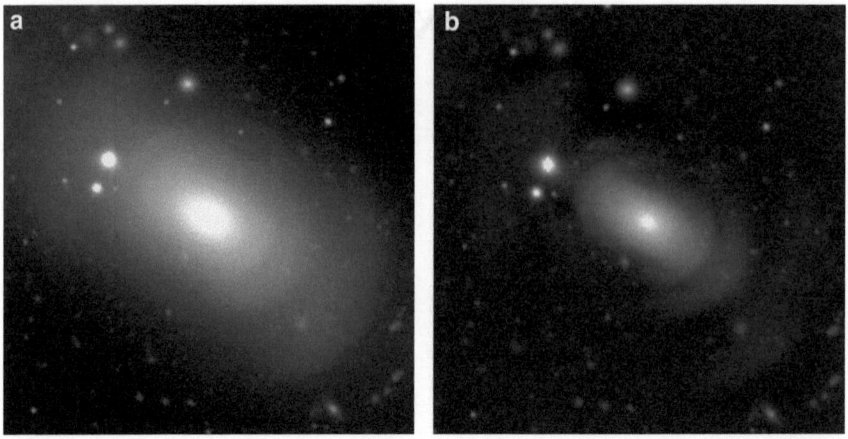

Fig. 1 Morphology of the central elliptical. Panel (**a**): VIMOS pre-image in the Bessel *R* band. Panel (**b**): The same frame after subtraction of a median box filter. Shells are clearly visible and show strong symmetry along the galaxy major axis, indicative of a recent minor merger. The field of view is ∼1.5 arcmin on a side

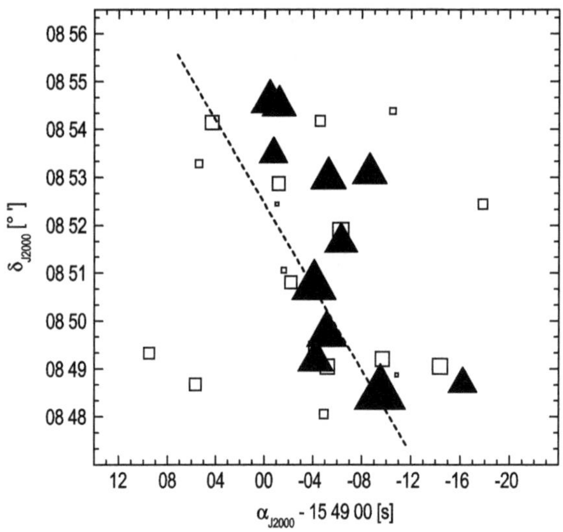

Fig. 2 Spatial distribution of spectroscopically determined RX J1548.9+0851 group members in the inner ∼300 kpc of the system. *Triangles* are objects older than 8 Gyr while *squares* represent younger objects. The symbol size scales with galaxy age. The *dashed line* indicates the orientation of the central elliptical major axis. Galaxies that formed their stars early are confined to an elongated structure aligned with the major axis of the central elliptical

have been determined in the surrounding faint galaxy population and fit by SSP models with the software package ULySS [2]. Figure 2 shows the distribution of the resulting SSP equivalent ages in the inner ∼300 kpc region of the system. There is a clear segregation in age for the investigated faint galaxy population.

While younger objects are diffusely distributed and found also in the outskirts of the system, the oldest objects (>8Gyr) form a central, elongated structure, aligned with the orientation of the major axis of the central elliptical.

Acknowledgements PE has been supported by the University of Vienna in the frame of the Initiativkolleg (IK) *The Cosmic Matter Circuit* I033-N.

References

1. L. Jones et al., MNRAS **343**, 627 (2003)
2. M. Koleva et al., A&A **501**, 1269 (2009)
3. W.A. Santos, C. Mendes de Oliveira, L. Sodré Jr., AJ **134**, 1551 (2007)

Measuring the Halo Mass Function in Loose Groups

D.J. Pisano, D.G. Barnes, B.K. Gibson, L. Staveley-Smith, K.C. Freeman, and V.A. Kilborn

Abstract Using data from our Parkes and ATCA HI survey of six groups analogous to the Local Group, we find that the HI mass function and velocity distribution function for loose groups are the same as those for the Local Group. Both mass functions confirm that the "missing satellite" problem exists in other galaxy groups.

1 Project Overview

Cold dark matter (CDM) models of galaxy formation predict that the Local Group should contain about 300 dark matter halos but there is an order of magnitude fewer galaxies observed [4, 5]. While the "missing satellite" problem can be mitigated by the inclusion of baryon physics in CDM models or alternate forms of dark matter, it is important to establish how this problem is manifest beyond the Local Group.

We have conducted a HI survey of six loose groups of galaxies that are analogous to the Local Group. The six groups are composed of only spiral and irregular galaxies that have mean separations of ~550 kpc. The groups have average

D.J. Pisano (✉)
Department of Physics, West Virginia University, P.O. Box 6315, Morgantown, WV 26506, USA
e-mail: djpisano@mail.wvu.edu

D.G. Barnes · V.A. Kilborn
Centre for Astrophysics and Supercomputing, Swinburne University, Hawthorn, VIC 3122, Australia

B.K. Gibson
Centre for Astrophysics, University of Central Lancashire, Preston, PR1 @HU, UK

L. Staveley-Smith
School of Physics, M013, University of Western Australia, Crawley, WA 6009, Australia

K.C. Freeman
RSAA, Mount Stromlo Observatory, Cotter Road, Weston, ACT 2611, Australia

I. Ferreras and A. Pasquali (eds.) *Environment and the Formation of Galaxies: 30 years later*, Astrophysics and Space Science Proceedings,
DOI 10.1007/978-3-642-20285-8_8, © Springer-Verlag Berlin Heidelberg 2011

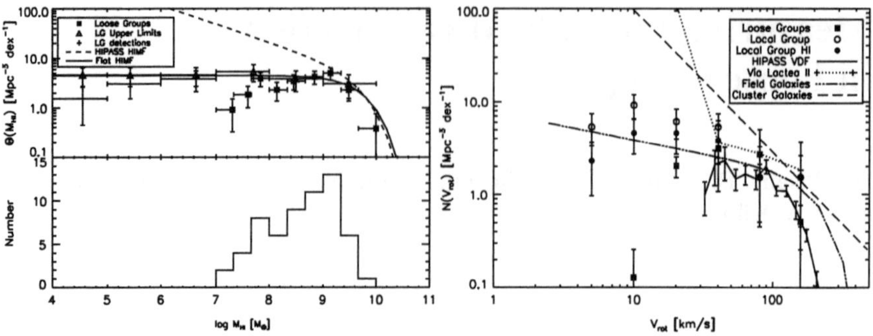

Fig. 1 *Left*: The HIMF for loose groups as compared to that for the Local Group galaxies with HI detections and Local Group galaxies with HI detections and upper limits. Also shown is the HIMF from HIPASS [7] and a flat HIMF. *Right*: The VDF for the loose groups compared to all Local Group galaxies, only those detected in HI, the HIPASS VDF from [7], field galaxies from [3], cluster galaxies from [1], and the theoretical predictions from Via Lactea II [2]. Aside from the loose and Local Group data, all other functions have been arbitrarily renormalized

diameters of 1.6 Mpc and have $M_{virial} \sim 10^{11.7-13.6}\,M_{\odot}$; they are similar to the Local Group in all these ways. Details on our observations, data reduction, and our search for HI clouds in the groups can be found in [6]. The survey identified a total of 63 group galaxies with all of the new detections having properties consistent with being typical dwarf irregular galaxies.

2 Halo Mass Functions

Using the survey completeness from [6] and our catalog of group galaxies, we constructed both a HI mass function (HIMF) and a circular velocity distribution function (VDF) for the six loose groups as shown in Fig. 1. The figure shows that both the HIMF and VDF for the Local Group are not atypical, but are consistent with those for the six loose groups. The HIMF for low density regions, such as loose groups (and including the Local Group), is consistent with being flatter than the HIMF in the field as was found by [7]. The VDF for loose groups has a consistent low mass slope to field galaxies and HIPASS galaxies [3, 8], but is much shallower than is predicted by simulations [2] or observed in galaxy clusters [1]. For a more complete discussion of these results, see Pisano et al. (2011, in preparation).

References

1. V. Desai et al., MNRAS **351**, 265 (2004)
2. J. Diemand et al., Nature **454**, 735 (2008)

3. A.H. Gonzalez et al., ApJ **528**, 145 (2000)
4. A. Klypin et al., ApJ **522**, 82 (1999)
5. B. Moore et al., ApJ **524**, L19 (1999)
6. D.J. Pisano et al., ApJ **662**, 959 (2007)
7. M.A. Zwaan et al., MNRAS **359**, L30 (2005)
8. M.A. Zwaan, M.J. Meyer, L. Staveley-Smith, MNRAS **403**, 1969 (2010)

The Environments of Luminous Infrared Galaxies

A.G. Tekola, P. Väisänen, and A. Berlind

Abstract This work studies the relationship between SFR, stellar mass and environment of local IR galaxies, especially LIRGs. It is found that the $L_{IR} \sim 10^{11} L_{\odot}$ marks an important luminosity point among IR galaxies - the large scale density of environment around LIRGs correlates strongly with their SFR above this value at a fixed stellar mass. On the other hand there is no clear correlation between stellar mass and density at constant SFR. No matter what the stellar mass is, local LIRGs in similar density environment tend to have the same SFR.

1 Introduction

Luminous infrared galaxies (LIRGs) have very high star formation rates (SFR) and are defined by having $10^{11} L_{\odot} \leq (L_{IR} = L_{8-1,000\,\mu\text{m}}) \leq 10^{12} L_{\odot}$. Their SF is often interaction triggered, a product of major mergers between gas rich disk-like galaxies. In the high-z Universe they tend to live in cluster environments [1]; however, their environment in the local Universe is not well defined. In this work [3] we aim at determining the dominant larger scale environment of local IR galaxies and LIRGs and to understand the relationships of SFR, environment, and stellar mass, with a sample drawn from the IRAS PSCz redshift survey. Environmental density estimates are made using 6dF galaxy counts around the IR galaxies within 2 Mpc radius and 10 Mpc length cylinder. SFR is derived from the L_{IR} values and

A.G. Tekola (✉) · P. Väisänen
South African Astronomical Observatory, P.O. Box 9, Observatory 7935, Cape Town, South Africa
e-mail: abiy@saao.ac.za; petri@saao.ac.za

A. Berlind
Department of Physics and astronomy, Vanderbilt University, Nashville, 1807 Station B, Nashville, TN 37235, USA
e-mail: a.berlind@vanderbilt.edu

I. Ferreras and A. Pasquali (eds.) *Environment and the Formation of Galaxies: 30 years later*, Astrophysics and Space Science Proceedings,
DOI 10.1007/978-3-642-20285-8_9,

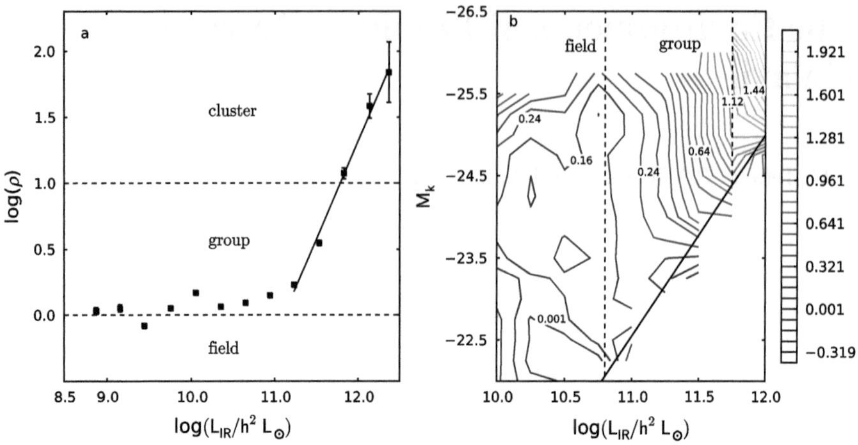

Fig. 1 *Left*: The relationship between L_{IR} and local density. *Right*: contour plot of M_K and L_{IR} at constant values of local density. The *black dashed lines* are boundaries between field/group (*both panels*) and group/cluster (*left panel*)

stellar mass is estimated from M_K available in 2MASS. Density boundaries between field/group/cluster are determined from virialized halo masses which the sample galaxies are associated to.

2 Results and Discussion

The left panel of Fig. 1 shows that the $L_{IR} \sim 10^{11} L_{\odot}$ luminosity point marks a break point with respect to environment among IR galaxies. Above this luminosity the SFR (or L_{IR}) of LIRGs and ULIRGs[1] is strongly correlated with environment. In terms of the traditional density classes (field, group and cluster), LIRGs live in groups while ULIRGs in our sample live in cluster environments. The fact that these galaxies dwell in high density 2 Mpc scale environments and that their SFR is correlated with their environment suggest that their SF mechanism is not triggered only by one-to-one interactions. Indeed, the tidal field of groups or clusters can enhance SFR beyond any one-to-one interactions [2]. The correlation between the L_{IR} and environment of our LIRGs also implies a SF-environment relationship similar to that found at $z \sim 1$ [1].

According to the right panel of Fig. 1 local LIRGs exhibit a correlation between their L_{IR} (a measure of SFR) and environment at constant M_K (or stellar mass), but they do not show any clear correlation between stellar mass and density at constant SFR. No matter what the stellar mass is, local LIRGs in similar density

[1]ULIRGs are defined as $L_{8-1,000\,\mu m} \geq 10^{12} L_{\odot}$.

environments tend to have the same SFR. The SF level of local LIRGs appears to be controlled by their environment regardless of their stellar mass. It seems that there is a minimum stellar mass threshold which galaxies have to have in order to be LIRGs at a certain environment and, above that stellar mass threshold, mass does not have any influence on their IR activity.

References

1. D. Elbaz et al., A&A **468**, 33 (2007)
2. M. Martig, F. Bournaud, MNRAS **385**, 38 (2008)
3. A.G. Tekola, P. Väisänen, A. Berlind, MNRAS (2011, in preparation) arXiv:1101.3495

UV–IR Luminosity Functions and Stellar Mass Functions of Galaxies in the Shapley Supercluster Core

A. Mercurio, C.P. Haines, P. Merluzzi, G. Busarello, R.J. Smith, S. Raychaudhury, and G.P. Smith

Abstract We present a panchromatic study of luminosity functions (LFs) and stellar mass functions (SMFs) of galaxies in the core of the Shapley supercluster at $z = 0.048$, in order to investigate how the dense environment affects the galaxy properties, such as star formation (SF) or stellar mass. We find that, while the faint-end slope of optical and NIR LFs steepens with decreasing density, no environment effect is found in the slope of the SMFs. This suggests that mechanisms transforming galaxies in different environments are mainly related to the quench of SF rather than to mass-loss. The Near-UV (NUV) and Far-UV (FUV) LFs obtained have steeper faint-end slopes than the local field population, while the 24 and 70 μm galaxy LFs for the Shapley supercluster have shapes fully consistent with those obtained for the local field galaxy population. This apparent lack of environmental dependence for the infrared (IR) LFs suggests that the bulk of the star-forming galaxies that make up the observed cluster IR LF have been recently accreted from the field and have yet to have their SF activity significantly affected by the cluster environment.

1 Introduction

Our panchromatic study of LFs in the Shapley supercluster core (SSC), from UV to IR wavebands, aims at investigating the relative importance of the processes that may be responsible for the galaxy transformations, examining, in particular, the

A. Mercurio (✉) · P. Merluzzi · G. Busarello
Istituto Nazionale di Astrofisica – Osservatorio Astronomico di Napoli, Naples, Italy
e-mail: mercurio@na.astro.it; merluzzi@na.astro.it; gianni@na.astro.it

C.P. Haines · S. Raychaudhury · G.P. Smith
School of Physics and Astronomy, University of Birmingham, Birmingham B15 2TT, UK

R.J. Smith
Department of Physics, University of Durham, Durham DH1 3LE, UK

I. Ferreras and A. Pasquali (eds.) *Environment and the Formation of Galaxies: 30 years later*, Astrophysics and Space Science Proceedings,
DOI 10.1007/978-3-642-20285-8_10,

effect of the environment through the comparison of LFs in regions with different local densities.

The SSC is constituted by three Abell clusters: A 3558, A 3562 and A 3556, and two poor clusters SC 1327-312 and SC 1329-313.

The target has been chosen since the most dramatic effects of environment on galaxy evolution should occur in superclusters, where the infall and encounter velocities of galaxies are greatest (>1,000 km s^{-1}), groups and clusters are still merging, and significant numbers of galaxies will be encountering the dense intra-cluster medium (ICM) of the supercluster environment for the first time. This work is carried out in the framework of the joint research programme ACCESS[1] (*A Complete Census of Star-formation and nuclear activity in the Shapley supercluster* [7]) aimed at determining the importance of cluster assembly processes in driving the evolution of galaxies as a function of galaxy mass and environment within the Shapley supercluster. We assume $\Omega_M = 0.3$, $\Omega_\Lambda = 0.7$ and $H_0 = 70\,\mathrm{km\,s^{-1}\,Mpc^{-1}}$.

2 UV–IR Luminosity Functions in Different Environments

We analysed wide field B- and R-band images aquired with the Wide Field Imager on the 2.2 MPG/ESO telescope (Shapley Optical Survey – SOS [6]) and K-band images obtained at the United Kingdom Infra-Red Telescope (UKIRT) with the Wide Field Infrared Camera (WFCAM), covering the whole of the SSC. These data are complemented by NUV and FUV GALEX data, and Spitzer/MIPS 24 μm and 70 μm imaging covering essentially the same region (∼2.0 deg^2) as the optical and NIR data. Full details of the observations, data reduction, and the production of the galaxy catalogues are described in [2, 6, 7] for the optical, NIR and UV/IR images, respectively.

We derived optical and NIR LFs in a 2.0 deg^2 area covering the SSC. In order to investigate the effects of the environment on galaxy properties, in Fig. 1 we show and compare the LFs in three different regions of the supercluster, characterised by high- ($\rho > 1.5$; filled circles, continuous lines), intermediate ($1.0 < \rho < 1.5$; open circles, long dashed lines), and low-density ($0.5 < \rho < 1.0\,\mathrm{gals\,arcmin^{-2}}$; crosses, short dashed) of galaxies. The local density was determined across the R-band WFI mosaic (see [6] for more details). The left and central panels show the B- and RS-band LFs, respectively.

The optical LFs cannot be described by a single Schechter function (S) due to the dip apparent at $M^{\star}+2$ both in B and R bands and the clear upturn in the counts for galaxies fainter than B and R ∼18 mag. Instead, the sum of a Gaussian and a Schechter function (G+S), for bright and faint galaxies, respectively, is a suitable representation of the data. Furthermore, the slope becomes significantly

[1] http://www.oacn.inaf.it/ACCESS.

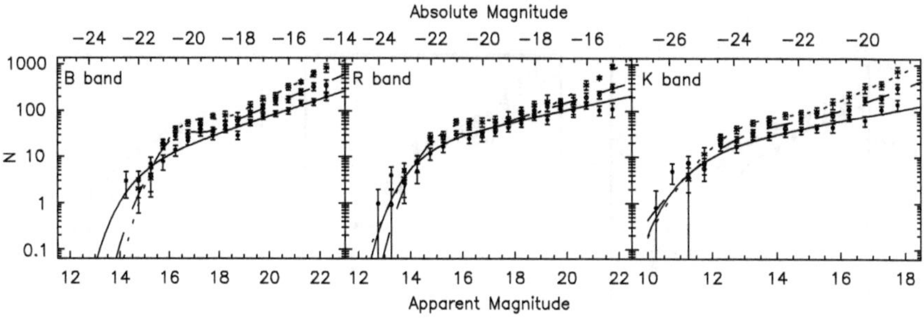

Fig. 1 Galaxies LFs B- (*left panel*), R-band (*central panel*) and K-band (*right panel*) high- (*filled circles*), intermediate- (*open circles*) and low-density (*crosses*) environments. The *long dashed* and *short dashed lines* represent the the G+S best fit for intermediate- and low-density environments respectively and the continuous line is the S best fit for high-density regions

steeper from high- to low-density environments, varying from −1.46±0.02 to −1.66±0.03 in B band and from −1.30±0.02 to −1.80±0.04 in R band, and being inconsistent at more than 3σ confidence level (c.l.) in both bands. Such a marked luminosity segregation is related to the behaviour of the red galaxy population: while red sequence counts are very similar to those obtained for the global galaxy population, the blue galaxy LFs are well described by single S and do not vary with the density (see [6]). This suggests that mechanisms transforming galaxies in different environments are mainly related to the quenching of SF.

In Fig. 1 (right panel), we show the K-band LFs of galaxies in the high- (filled circles), intermediate- (open circles) and low-density (crosses) regions together with their best fits (see figure's caption). Although the fit with a single S function cannot be rejected in all the three environments, the LFs suggest a bimodal behaviour due to the presence of an upturn for faint galaxies. To successfully model these changes in slope and to compare our results with our optical LFs, we fit our data with a composite G+S. Moreover, the faint-end slope becomes steeper from high- to low-density environments varying from −1.33±0.03 to −1.49±0.04, being inconsistent at the 2σ c.l. between high- and low-density regions. The observed trend with environment confirms those observed for the Shapley optical LFs although at a lower significance level.

From the GALEX data we obtained NUV and FUV LFs (Fig. 2, left panel), which were found to have steeper faint- end slopes ($\alpha = -1.5 \pm 0.1$) than the local field population ($\alpha = -1.2 \pm 0.1$) at the $\sim 2\sigma$ level, largely due to the contribution of massive, quiescent galaxies at $M_{FUV} \sim -16$. Using the Spitzer/MIPS 24 μm imaging, we determined the IR LF of the SSC, finding it well described by a single Schechter function with $\log(L^*_{IR}/L_\odot) = 10.52^{+0.06}_{-0.08}$ and $\alpha_{IR} = -1.49 \pm 0.04$. We also presented the first 70 μm LF of a local cluster with Spitzer, finding it to be consistent with that obtained at 24 μm. The shapes of the 24 and 70 μm LFs were also found to be indistinguishable from those of the local field population (grey squares in Fig. 2 central and right panel). This apparent lack of environmental

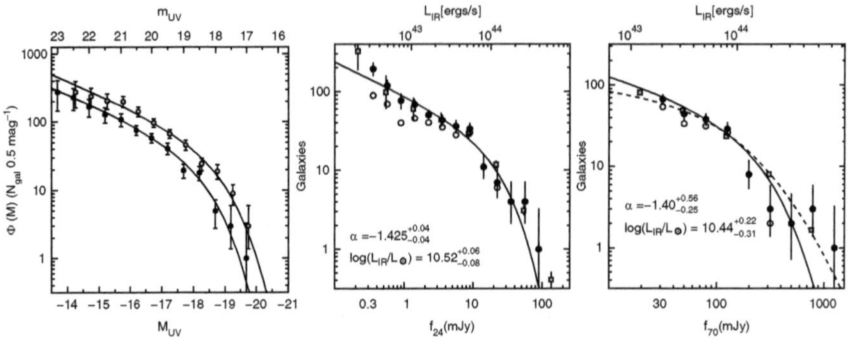

Fig. 2 *Left panel* FUV (*black*) and NUV (*grey*) LFs. The *solid lines* indicate the best-fitting S functions to the data. *Central and right panel:* 24 and 70 μm LF (*solid black symbols*), respectively. In both panels the contribution due to spectroscopically confirmed supercluster members is indicated by *open symbols* and the best fitting S is indicated by *solid black curves*. *Grey squares* represent the the field 24 μm LF of [5] in the *central panel* and in the *right panel* the field LF of IRAS galaxies from [9], while the *dashed curve* indicates the analytic form of the IRAS field IR LF of [8]

dependence for the shape of the FIR luminosity function suggests that the bulk of the star-forming galaxies that make up the observed cluster infrared LF have been recently accreted from the field and have yet to have their SF activity significantly affected by the cluster environment. As the SF is quenched via cluster-related processes, the UV and IR emission drops rapidly, and reduce the normalization factor of the corresponding LFs without affecting their shape.

3 The Galaxy Stellar Mass Functions

The combined optical and NIR data allow us to derive the distribution of galaxy stellar masses (SMF). The sample we analysed is in the magnitude range $10 \leq K \leq 18$ and refers to the $\sim$2 deg^2 area covered by both the SOS and our K-band imaging. The stellar masses of galaxies belonging to the SSC are estimated by means of stellar population models by Maraston [4] with a Salpeter initial mass function constrained by the observed optical and infrared colours (see [7] for details). We use the probability that galaxies are supercluster members as derived by [1] following [3] in order to estimate and correct for the foreground/background contamination. We choose that stellar masses of galaxies belonging to the Shapley supercluster contribute to the galaxy SMF according to their likelihood of belonging to the SSC.

In Fig. 3 we show the SMF for the different supercluster environments. Unlike in the case of the optical and NIR LFs no environmental trend is seen in the slope of the

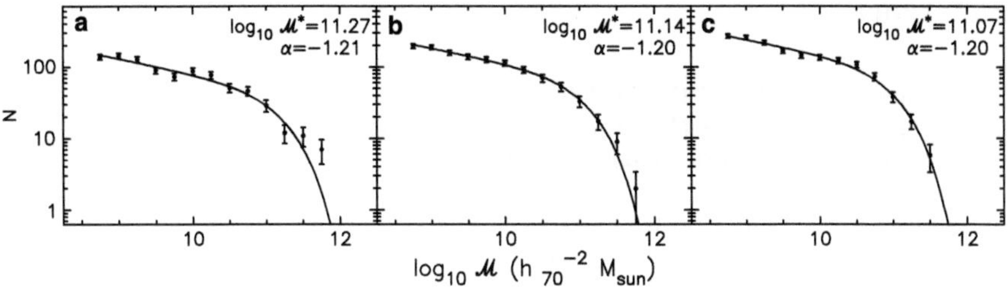

Fig. 3 The stellar mass function of galaxies in the three cluster regions corresponding to high- (panel **a**), intermediate- (panel **b**) and low-density (panel **c**) environments. In the *left*, *central* and *right panel* the *solid line* represents the fit to the data. In each panel the best fit value of α and $\log_{10} \mathcal{M}^{\star}$ are reported

SMFs. On the other hand, the increase in $\mathcal{M}^{\star}$ from low- to high-density regions and the excess of high-mass galaxies depend on environment. The different behaviours of LF and of galaxy SMF with the environment confirm that the mechanisms transforming galaxies in different environments are mainly related to the quenching of SF rather than to mass-loss.

4 Conclusions

We find that the faint-end slope of the optical and K-band LFs becomes steeper from high- to low-density environments, although this change is less dramatic at NIR wavebands, indicating that the faint galaxy population increases in low-density environments. Differently from the LF, no environmental effect is found in the slope of the SMFs. On the other hand, the increase in $\mathcal{M}^{\star}$ from low- to high-density regions and the excess of galaxies at the bright-end are also dependent on environment. These results seem to indicate that the physical mechanisms responsible for the transformation of galaxies properties in different environments are mainly related to the quenching of the SF. Moreover, the NUV and FUV LFs have steeper faint-end slopes than the local field population, while the 24 and 70 μm galaxy LFs for the Shapley supercluster have shapes fully consistent with those obtained for the Coma cluster and for the local field galaxy population. This apparent lack of environmental dependence for the shape of IR luminosity functions suggests that the bulk of the star-forming galaxies that make up the observed cluster infrared LF have been recently accreted from the field and have yet to have their SF activity significantly affected by the cluster environment.

Acknowledgements This work was carried out in the framework of the FP7-PEOPLE-IRSES-2008 project ACCESS. AM acknowledges financial support from INAF-OAC and the JENAM grant to attend the conference.

References

1. C.P. Haines et al., MNRAS **371**, 55 (2006)
2. C.P. Haines et al., MNRAS **412**, 127 (2011). arXiv:1010.4323
3. T. Kodama, R. Bower, MNRAS **321**, 18 (2001)
4. C. Maraston, MNRAS **362**, 799 (2005)
5. F.R. Marleau et al., ApJ **633**, 218 (2007)
6. A. Mercurio et al., MNRAS **368**, 109 (2006)
7. P. Merluzzi et al., MNRAS **402**, 753 (2010)
8. T.T. Takeuchi et al., ApJL **587**, 89 (2003)
9. L. Wang, M. Rowan-Robinson, MNRAS **401**, 35 (2010)

Component Luminosity Functions of Galaxy Pairs in the MGC

K. Casteels and D. Patton

Abstract We present new techniques for obtaining luminosity functions (LFs) of galaxies in close pairs as a probe for luminosity changes in interacting and merging systems (Casteels and Patton 2011). Using mock catalogues, the effect of imposing a fixed magnitude separation (ΔM) between galaxy pairs (to isolate major mergers) is explored and we find that imposing a ΔM introduces a selection bias which must be corrected for when comparing with field LFs. Our method is then applied to the Millennium Galaxy Catalogue (MGC) to obtain the LF of galaxies in pairs for $\Delta M = 6$ and $\Delta M = 1$ with $r_p < 50\,\text{kpc}$, $\Delta V < 500\,\text{km s}^{-1}$, $-16 > M_B > -22$ and $0.013 < z < 0.18$. We find that the global LF and the LF of galaxies in pairs gain power towards the bright end compared to their corresponding field LF. The major merger ($\Delta M = 1$) pairs LFs all gain additional power on the bright end relative to the major + minor ($\Delta M = 6$) pairs LFs, indicating that the brightest galaxies are favoured in major mergers. Bulges and ellipticals are also found to be strongly favoured over discs in close pairs.

1 Pairs Luminosity Function Techniques and the MGC

Several weights must be applied when creating luminosity functions (LFs) of galaxies in close pairs. To test various weighting schemes we generated 10 mock catalogues with 100,000 galaxies each, with parameters matching the MGC field

K. Casteels (✉)
Departament d'Astronomia i Meteorologia, Universitat de Barcelona, Martí Franques 1, 08028 Barcelona, Spain
e-mail: kcasteels@am.ub.es

D. Patton
Department of Physics and Astronomy, Trent University, 1600 West Bank Drive, Peterborough, ON, Canada K9J 7B8
e-mail: dpatton@trentu.ca

I. Ferreras and A. Pasquali (eds.) *Environment and the Formation of Galaxies: 30 years later*, Astrophysics and Space Science Proceedings,
DOI 10.1007/978-3-642-20285-8_11, © Springer-Verlag Berlin Heidelberg 2011

LF results of [4]. Bulge and disc components were generated by simply dividing a galaxy's total luminosity in two. We find:

1. For a flux limited sample, the Vmax weight of the *faint* galaxy ($w_{Vmax_{faint}}$) must be applied to both galaxies in a pair. In previous pairs LF studies by [6–9], a galaxy's individual Vmax was used, while [10] and [4] used the Vmax of the brighter galaxy for both. If the faint galaxy is not observed, neither is the pair. Bulges and discs should be treated separately from their host galaxies, with the largest Vmax weight between their host galaxies applied to both.
2. When applying a maximum ΔM between pairs to isolate major mergers, we find that as ΔM decreases, the resulting pairs LF is skewed towards magnitudes with higher number densities. To correct this selection bias we apply a weight to each paired galaxy equal to the number of galaxies in the sample divided by the number of galaxies within ΔM around it:

$$w_{\Delta M} = \int_{M_{bright}}^{M_{faint}} \phi_{field}(M) / \int_{M-\Delta M}^{M+\Delta M} \phi_{field}(M), \tag{1}$$

where ϕ_{field} is the field LF, M_{faint} and M_{bright} are the sample magnitude limits, and M is the absolute magnitude of the paired galaxy in question. The same $w_{\Delta M}$ is applied to bulges and discs as for the global fit since their probability of being observed is determined by their host galaxy.

Following the methods outlined in [5] weights should also be applied to account for pairs near the survey boundaries (w_{b_2}), pairs near the redshift boundaries (w_{v_2}), global spectroscopic incompleteness weights (w_s), as well angular spectroscopic incompleteness weights to correct for fibre collisions at small angular separations (w_{θ_2}).

These techniques are applied to the MGC GIM2D bulge-disc decomposition catalogue (mgc-gim2d.cat) which [1] describe in detail. Results are obtained for pairs with $r_p < 50\,h^{-1}$ kpc and $\Delta V < 500\,h^{-1}$km s^{-1} between $-16 > M_B > -22$ and $0.013 < z < 0.18$. Our final sample contains 6143 galaxies, with 775 in pairs for $\Delta M < 6$ and 412 for $\Delta M < 1$. We find:

1. All of the pairs LFs gain power towards the bright end compared to their corresponding field LF, which is indicative of strong luminosity dependent clustering.
2. The major mergers pairs LFs ($\Delta M < 1$) all gain power on the bright end compared to $\Delta M < 6$ pairs, indicating that major mergers are most common amongst the brightest galaxies.
3. Bulges and ellipticals are more common in close pairs than discs, indicating that discs are not favoured in close galaxy interactions.

Please refer to [2] for a complete description of the techniques described here and a more thorough discussion of the results obtained for the MGC.

References

1. P.D. Allen et al., MNRAS **371**, 2 (2006)
2. K.R.V. Casteels, D.R. Patton, AA (2011, in preparation)
3. D.L. Domingue et al., ApJ **695**, 1559 (2009)
4. S.P. Driver et al., ApJL **657**, L85 (2007)
5. D.R. Patton et al., ApJ **565**, 208 (2002)
6. D.S.L. Soares et al., AAPS **110**, 371 (1995)
7. J.W. Sulentic, C.R. Rabaca, ApJ **429**, 531 (1994)
8. H.M.H. Toledo et al., AJ **118**, 108 (1999)
9. C. Xu, J.W. Sulentic, ApJ **374**, 407 (1991)
10. C.K. Xu et al., ApJL **603**, L73 (2004)

The Hercules Cluster Environment Impact on the Chemical History of Star-Forming Galaxies

V. Petropoulou, J.M. Vílchez, J. Iglesias-Páramo, and P. Papaderos

Abstract In this work we study the effects of the Hercules cluster environment on the chemical history of star-forming (SF) galaxies. For this purpose we have derived the gas metallicities, the mean stellar metallicities and ages, the masses and the luminosities of our sample of galaxies. We have found that our Hercules SF galaxies are either chemically evolved spirals with nearly flat oxygen gradients, or less metal-rich dwarf galaxies which appear to be the "newcomers" in the cluster. Most Hercules SF galaxies follow well defined mass-metallicity and luminosity-metallicity sequences; nevertheless significant outliers to these relations have been identified, illustrating how environmental effects can provide a physical source of dispersion in these fundamental relations.

1 Mass vs. Metallicity and Metallicity vs. Local Density

The Hercules cluster is one of the most exciting nearby dense environments in the local Universe, showing abundant sub-structures unraveled in X-ray emission and broadband imaging. This cluster appears to be currently assembled via the merger of smaller substructures, a process that can trigger the extraordinary interaction and merging activity observed and constitutes an ideal laboratory to explore the effects of environment on galaxy evolution.

In this work we have investigated the impact of the environment of the Hercules cluster on the chemical content of a sample of 31 SF galaxies for which deep

V. Petropoulou (✉) · J.M. Vílchez · J. Iglesias-Páramo
Instituto de Astrofísica de Andalucía-CSIC, Glorieta de la Astronomía, 18008, Granada, Spain
e-mail: vasiliki@iaa.es; jvm@iaa.es; jiglesia@iaa.es

P. Papaderos
Centro de Astrofísica da Universidade do Porto, Rua das Estrelas, 4150-762, Porto, Portugal
e-mail: papaderos@astro.up.pt

I. Ferreras and A. Pasquali (eds.) *Environment and the Formation of Galaxies: 30 years later*, Astrophysics and Space Science Proceedings,
DOI 10.1007/978-3-642-20285-8_12, © Springer-Verlag Berlin Heidelberg 2011

Hα imaging was previously obtained [1]. We have obtained new long-slit optical spectra for 27 of them, with ISIS/WHT and IDS/INT in order to perform a spatially resolved study of the effects of environment on these galaxies. Spectral synthesis model fitting has been performed for all the spectra analyzed in order to provide an effective correction for the underlying stellar absorption on emission line spectra, as well as to derive the characteristic properties of the galaxy stellar populations. Line fluxes and chemical abundances of O/H and N/O have been obtained for all the galaxies, and, whenever possible, for different parts of galaxies of the sample.

We have found that our sample galaxies shows well defined sequences of blue luminosity vs. metallicity and stellar mass vs. metallicity (using both O/H and also the N/O ratio), following the general behavior found for SF galaxies. This result goes in the line of most recent findings showing that these global relations seem not to be significantly affected by environment [2, 3]. Besides this global behavior, we have found that a set of galaxies, previously identified as environmentally affected, appear clearly to fall off the main global sequences, thus pointing to environment as the physical source for the dispersion observed in these relations.

From the study of the metallicity vs. galaxy local density we have seen a dual behavior separating the dwarfs from the more luminous galaxies. The latter galaxies, populating the cluster dense core (up to log $\Sigma_{4,5} \sim 2.5$), do not show any significant trend with metallicity, whereas the observed less metal-rich dwarfs are found preferentially at lower densities and seem to be the "newcomers" to the cluster (Fig. 1).

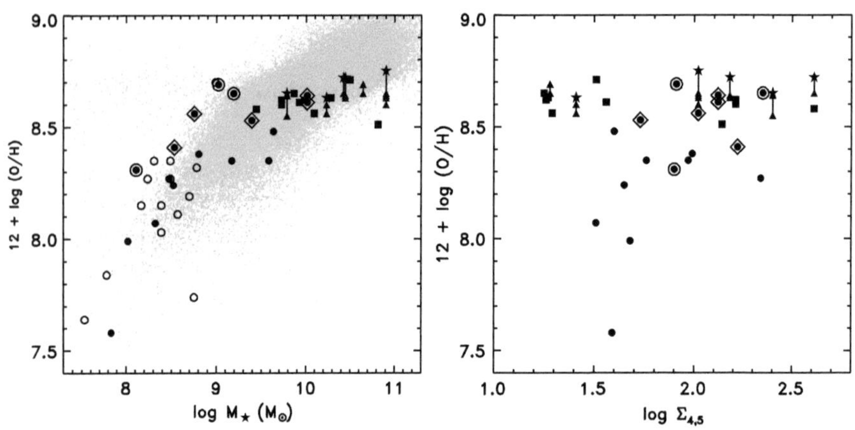

Fig. 1 *Left*: Galaxy stellar mass vs gas-phase oxygen abundance for the Hercules galaxies. We use *filled circles* to indicate dwarf/irregular galaxies ($B > -19$), *squares* for spirals ($B < -19$) but integrated galaxies, *stars* for the nuclei of six spirals that we divide into different parts and *triangles* for their corresponding disk components (nuclei and disks are connected with *lines*). We mark with *circles* the Tidal Dwarf Candidates of our sample and with diamonds the dwarfs with close companions. *Open circles* correspond to Virgo dIs and BCDs from [4] and the gray points to MPA/JHU SF galaxies. *Right*: Oxygen abundance versus the local galaxy number density to the average of the projected distances to the fourth and fifth nearest neighbor $\Sigma_{4,5}$, where the galaxies considered are secure members of the Hercules cluster, as follows by their SDSS spectroscopic redshift

References

1. B. Cedrés et al., AJ **138**, 873 (2009)
2. S.L. Ellison et al., MNRAS **396**, 1257 (2009)
3. M. Mouhcine, I.K. Baldry, S.P. Bamford, MNRAS **382**, 801 (2007)
4. O. Vaduvescu, M.L. McCall, M.G. Richer, AJ **134**, 604 (2007)

The Symbiotic Relationship Between the Evolution of Galaxy Groups and Their Resident Galaxies

E.M. Wilcots

Abstract The vast majority of galaxies reside in groups and it is becoming increasingly clear that much of the transformation of galaxies occurs in this intermediate environment rather than in larger clusters. We report on observational and analytical work that shows how the evolution of the group environment influences the evolution of the resident galaxies and how feedback resulting from the evolution of individual galaxies influences their larger environment. In particular, the dynamical assembly of groups is marked by an increase in galaxy–galaxy interactions and mergers that ultimately result in a group with a significant population of elliptical galaxies and a diffuse X-ray halo. The infall of baryons into the group produces a hot intragroup medium whose temperature can be maintained by episodic AGN feedback.

1 Why Galaxy Groups?

It is now well established that not only do a vast majority of galaxies reside in groups, but that individual galaxies also spend a large fraction of their lifetimes in the group environment. Given this, it is important to ask both how the evolution of the group impacts the evolution of individual galaxies and how the evolution of the resident galaxies influences the group environment.

We have seen a number of attempts to quantitatively define what a "group" actually is. At one limit there are a number of "poor clusters" that have about 30 members with apparent magnitude, m, such that $m_3 < m < (m_3+2)$ where m_3 is the magnitude of the 3rd brightest member [1]. More recent approaches define groups as having 3-10 members with $M_V < -19$ within a radius of 0.5 Mpc [11] or with

E.M. Wilcots (✉)
University of Wisconsin-Madison, 475 N. Charter St. Madison, WI 53706, USA
e-mail: ewilcots@astro.wisc.edu

I. Ferreras and A. Pasquali (eds.) *Environment and the Formation of Galaxies: 30 years later*, Astrophysics and Space Science Proceedings,
DOI 10.1007/978-3-642-20285-8_13, © Springer-Verlag Berlin Heidelberg 2011

having a few members with a velocity dispersion of $200\,\mathrm{km\,s^{-1}} < \sigma < 400\,\mathrm{km\,s^{-1}}$ [4]. In the context of modern cosmological simulations, groups are typically defined as having halo masses of $10^{12} - 10^{14}\,M_{\odot}$ [6]. Observationally, one can use a friends-of-friends algorithm to extract large samples of galaxy groups from redshift surveys [3].

It is important to recognize that groups are not simply scaled down versions of clusters. There are quantitative differences between the galaxy luminosity function, H I mass function, and X-ray scaling relationships between groups and clusters that imply that the galaxy population in groups is not the same as the galaxy population in clusters [6,7]. Perhaps the most fundamental difference between these two environments is that the velocity dispersion in a group is comparable to the internal velocities of the resident galaxies. Galaxy–galaxy interactions and mergers are much more common in groups than in the field where the velocity dispersion is significantly higher.

2 Dynamical History of Groups

The history of a group is perhaps better described in terms of the timescale of mass assembly. That is, how long ago did the galaxies in the group become gravitationally bound? In an investigation of large-scale cosmological simulations, [5] identified "evolved" groups as those with assembly times of 5–8 Gyr and that are now dynamically relaxed, X-ray luminous, and dominated by elliptical galaxies. "Young" groups, on the other hand, may still be collapsing, lack luminous X-ray emission, and are spiral rich [5, 8].

In a study of the neutral gas content of galaxy groups spanning a range of dynamical states, [7] showed that galaxy–galaxy interactions were a common feature of intermediate age groups. In this case "intermediate age" refers to groups whose galaxy population is mixed, implying some dynamical evolution while still maintaining a population of gas-rich spirals. What [7] find in their survey is that nearly every gas rich galaxy in such groups is either interacting with another galaxy or shows clear signs of having suffered a recent interaction. The implication of these results is that galaxy–galaxy interactions are the primary driver of galaxy transformation in groups.

NGC 2563 is one the most well-studied dynamically evolved groups, having 63 members within a radius of 1.15 Mpc and a velocity dispersion of $\sigma = 364\,\mathrm{km\,s^{-1}}$ [2]. In [2] we see that the group has a large X-ray halo centered on the massive elliptical NGC 2563 while an H I mosaic carried out with the Very Large Array shows that all of the H I detected galaxies reside beyond the extent of the X-ray halo but in a very asymmetric distribution as seen in Fig. 1 [2]. We see a similar trend in the H I distribution in another evolved group, NGC 5846.

Synthesizing the work of [7] and the recent detailed study of NGC 2563 suggests that the dynamical evolution of galaxy groups is marked by a phase in

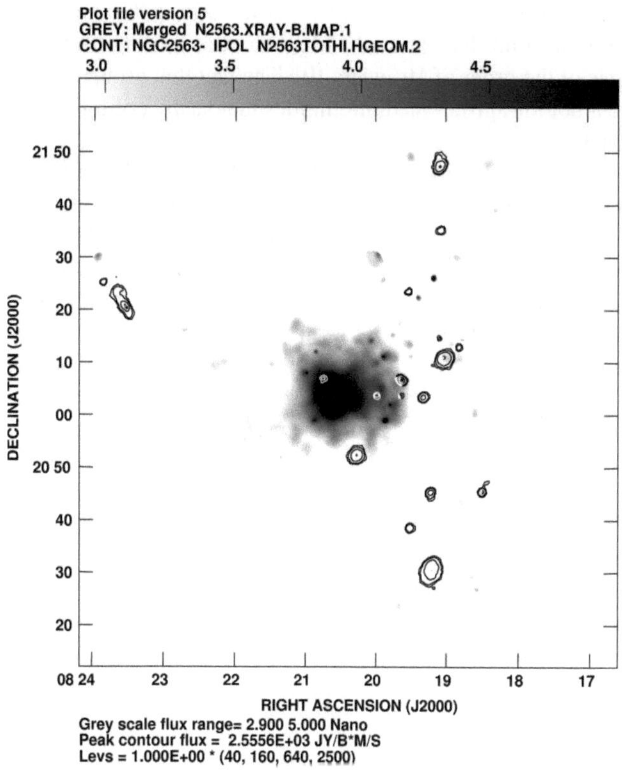

Fig. 1 This figure shows the spatial relationship between H I detected galaxies, which are shown as contours, and the extended diffuse X-ray halo, which is shown as a greyscale, in the evolved galaxy group NGC 2563 [2]

which interactions between gas-rich spiral galaxies are common, leaving behind a gravitationally bound system dominated by elliptical galaxies and having a large X-ray halo. Further infall of gas rich galaxies onto such an evolved halo continues the growth of the group as we see in the population of gas-rich galaxies on the outskirts of NGC 2563. Most importantly, we see that the dynamical evolution of galaxy groups drives both the transformation of galaxies through galaxy mergers and acquisitions, and the transformation of the intergalactic medium.

3 Transformation of the Intragroup Medium

The evolution of the group is reflected in changes in its overall baryonic content. Baryons in the intergalactic medium will be accreted into a group with infall velocities that far exceed the speed of sound, resulting in the shock heating of gas to high temperatures. Under most conditions, the gas should experience catastrophic cooling on timescales less than a Hubble time, particularly in the inner parts of the group. Analytically, we find that for a $10^{12}\,M_{\odot}$ halo, the cooling time in the inner part of the group is of the order of 10^9 years [9]. The fact that groups in general are observed to have a hot intragroup medium implies that such cooling should trigger black hole feedback.

In [9] we show that AGN feedback could episodically reheat the baryonic content of galaxy groups on timescales that are linked to the cooling time. The initial temperature of the intergalactic baryons should be a result of the shock-heating of the gas during infall. How the temperature falls off with radius within the group will depend on the radial density distribution. Our simulations assume a r^{-3} and a $r^{-3/2}$ density distribution, with the AGN turning on after t_{cool} and shutting off at 10^8 years with an efficiency of 10% (a measure of how efficiently that AGN heats the intergalactic medium). The intergalactic gas then cools and the process repeats itself. Our simulations thus suggest a symbiotic relationship between the evolution of the baryonic content of groups and the evolution of the resident galaxies.

We do not yet know how much energy AGN deposit into the intragroup medium. In these models we have assumed that the AGN is triggered by cooling of the IGM on timescales comparable to the cooling time. With a duty cycle of 5×10^8 years, a jet luminosity of 10^{43} erg s^{-1}, and efficiencies of only $\sim$0.1%, jets can impart $\sim$1 keV per particle over the virial radius of the group [9]. With such conservative estimate radio jets can certainly alter the thermodynamics of the baryonic content of groups quite dramatically.

4 A Statistical Challenge

Fundamentally, understanding the influence of AGN feedback on the evolution of galaxy groups comes down to knowing the frequency with which AGN occur in these intermediate environments. With the number of large catalogs of galaxy groups now being culled from large-scale surveys, we now have the opportunity to explore this question. The 2dF Galaxy group catalog [6], for example, contains 2×10^5 groups culled from the 2dF Galaxy Redshift survey that covered 15,000 square degrees and lists 2.1×10^6 galaxies. Sadler et al. [10] cross-correlated the 2dFGRS galaxies with radio sources in the NRAO VLA Sky Survey. This yielded $\sim$4,000 matched sources, 2,200 of which could be classified as AGN with a radio flux of $\geq$ 50 mJy [10]. We found that 70% of radio-loud AGN reside in 2dF groups as opposed to 55% of "normal" galaxies. More interestingly, only 8% of groups with more than four members host a radio-loud AGN. This amounts to 4.5% of all groups, and extremely few groups have more than one radio-loud AGN.

This preliminary study is limited to only radio-loud sources; the fraction of groups hosting an AGN will certainly be larger with a lower cut-off on the radio flux limit. The new generation of radio telescopes offers the opportunity to probe the population of the lower luminosity radio sources in groups. Alternatively, we can look at the population of nearby groups and study the population of low luminosity

AGN in these systems. We have begun to do this by looking at Seyfert galaxies in groups in the Coma-Abell 1367 supercluster. It turns out that half of the groups in this supercluster contain at least one Seyfert galaxy. Most of these are likely the result of the galaxy–galaxy interactions that are common in the group environment.

5 Summary and Conclusions

Galaxy groups are proving to be complex environments in which the morphology of individual galaxies and the nature of the intergalactic medium are linked to the dynamical evolution of the groups themselves. The assembly of groups affects morphology and the gas content of individual galaxies largely through galaxy–galaxy mergers and acquisitions. Along the way the infall of baryons onto the group and their subsequent cooling may be linked to the onset of AGN activity in the group's central galaxy. The feedback from this activity affects the thermodynamics of the IGM. The influence of the IGM on the morphology and gas content of galaxies in groups remains to be seen.

References

1. J.N. Bahcall, ApJ **238L**, 117 (1980)
2. X.N. Bai et al., (2010, in preparation)
3. A. Berlind et al., ApJS **167**, 1 (2006)
4. R.G. Carlberg et al., ApJ **552**, 427 (2001)
5. A. Dariush et al., MNRAS **405**, 1873 (2010)
6. V.R. Eke et al., MNRAS **355**, 769 (2004)
7. E. Freeland, A. Stilp, E. Wilcots, AJ **138**, 295 (2009)
8. J. Rasmussen et al., MNRAS **373**, 653 (2006)
9. M. Riabokin, E.M. Wilcots, ApJ (2010, submitted)
10. E. Sadler et al., MNRAS **329**, 227 (2002)
11. E.L. Zirbel, ApJ **476**, 489 (1997)

Lopsidedness in WHISP Galaxies

E. Jütte, J. van Eymeren, Ch. Jog, R.-J. Dettmar, and Y. Stein

Abstract Observations of the stellar and gaseous components in disc galaxies often reveal asymmetries in the morphological and kinematic distribution. However, the origin of this effect is not well known to date, and quantitative studies are rare. Here, we present the first statistical investigation of a sample of 76 HI discs using the WHISP survey. We perform a Fourier analysis to study the morphological lopsidedness. This allows to trace the degree of asymmetry with radius. We further investigate the dependence on, e.g., the morphological type and environment.

1 Introduction

It has been known for a long time that there are non-axisymmetric characteristics in discs of spiral galaxies in the stellar and gaseous morphology and/or kinematics and therefore in the mass distribution (e.g., [2, 6, 9]). Lopsided galaxies show a global asymmetry so that the mass distribution can be characterised by the Fourier amplitude $m = 1$. Lopsidedness is particularly seen in the outer optical discs of 50% of the galaxies (e.g., [8,9]). Most of the studies, however, are based on optical imaging or global HI profiles. HI maps are ideal to investigate lopsidedness since gas discs are much further extended than stellar discs. Also, both morphological and kinematic information are provided at the same time.

E. Jütte (✉) · R.-J. Dettmar · Y. Stein
Astronomisches Institut der Ruhr–Universität Bochum (AIRUB), Bochum, Germany
e-mail: eva.juette@astro.rub.de

J. van Eymeren
Fakultät für Physik, Universität Duisburg-Essen, Essen, Germany

Ch. Jog
Department of Physics, Indian Institute of Science, Bangalore, India

I. Ferreras and A. Pasquali (eds.) *Environment and the Formation of Galaxies: 30 years later*, Astrophysics and Space Science Proceedings,
DOI 10.1007/978-3-642-20285-8_14, © Springer-Verlag Berlin Heidelberg 2011

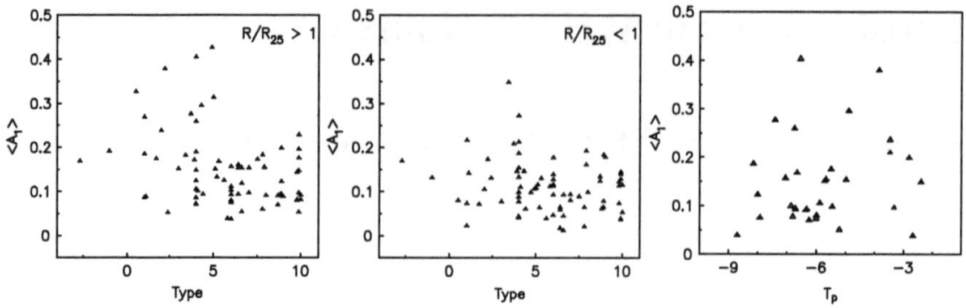

Fig. 1 *Left*: $\langle A_1 \rangle$ vs. morphological type averaged over large and small radii. *Right*: $\langle A_1 \rangle$ vs. tidal parameter

Possible scenarios for lopsidedness are tidal interaction [5] or minor merger [10], continuous gas accretion [3], but also an offset of the stellar disc in a halo potential [7].

We used a sub-sample of the WHISP[1] survey to systematically analyse a large set of HI discs with respect to the local and global morphology. Our sample includes 76 galaxies which were chosen according to the following selection criteria: (1) the ratio of the HI diameter over the beam size has to be larger than 10 at a resolution of 30 arcsec; (2) the galaxies need to have inclinations between 20 and 75°. This sample covers a whole range of morphological types and galaxy masses. We trace the neutral gas out to at least twice the optical radius, sometimes even further.

2 Method and Results

Lopsidedness can be described as $m = 1$ mode of a Fourier decomposition [9]:

$$\sigma(r, \phi) = a_0(r) + \sum a_m(r) \cdot cos[m \cdot \phi' - \phi_m(r)] \qquad (1)$$

Here, a_0 is the mean surface density, a_m the amplitude of the surface density harmonic coefficient, ϕ_m the phase and ϕ' the azimuthal angle in the plane of the galaxy. Lopsidedness can then be characterised by the normalised Fourier amplitude $A_1 = a_1/a_0$ and the phase ϕ_1. We performed a harmonic decomposition of the HI intensity maps using the kinematic parameters derived from a tilted-ring analysis.

We found that within the optical disc A_1 increases continuously with radius, but seems to saturate beyond that. Early-type disc galaxies tend to show a higher lopsidedness than late-type galaxies, in particular in the outer parts (Fig. 1, left panels). This is in agreement with [1]. We calculated the tidal parameter in order to investigate the correlation of lopsidedness with the environment [3]. As Fig. 1,

[1] Westerbork HI Survey of Irregular and Spiral galaxies.

right panel shows, lopsidedness and the environment of the sample galaxies are not correlated (in agreement with [3]). Both results together indicate a tidal origin for the lopsidedness: the disc responds to a halo that is distorted by a tidal encounter [5]. For a detailed analysis see [4].

References

1. R.A. Angiras et al., MNRAS **369**, 1849 (2006)
2. J.E. Baldwin, D. Lynden-Bell, R. Sancisi, MNRAS **193**, 313 (1980)
3. F. Bournaud et al., A&A **438**, 507 (2005)
4. J. van Eymeren et al., A&A **530**, id.A29 (2011, submitted)
5. C.J. Jog, ApJ **488**, 642 (1997)
6. C.J. Jog, F. Combes, Phys. Rep. **471**, 75 (2009)
7. E. Noordermeer, L.S. Sparke, S.E. Levine, MNRAS **328**, 1064 (2001)
8. O. Richter, R. Sancisi, A&A **290**, L9 (1994)
9. H.-W. Rix, D. Zaritsky, D., ApJ **447**, 82 (1995)
10. D. Zaritsky, H.-W. Rix, ApJ **477**, 118 (1997)

The Fundamental Plane of Early-Type Galaxies: Environmental Dependence from *g* Through *K*

F. La Barbera, P.A.A. Lopes, and R.R. de Carvalho

Abstract We analyze the dependence of the Fundamental Plane (FP) relation on the environment where galaxies reside, for a sample of $39,993$ early-type galaxies (ETGs) in the optical (griz) and Near-Infrared (YJHK) wavebands. The intercept "c" of the FP decreases from high- to low-density regions, implying that galaxies at low density have on average lower mass-to-light ratios than their high-density counterparts. Since the variation of "c" is the same at all wavebands, we conclude that ETGs at low density have younger luminosity-weighted ages (and higher metallicity) than cluster ETGs. The velocity dispersion slope of the FP, "a", is smaller for groups relative to field ETGs, independent of the waveband. The surface brightness slope, "b", does not change with waveband for group galaxies, while exhibits a small, but significant increase from g through K for the field sample. We interpret these trends as the result of a different variation of dark-matter fraction and stellar population properties with galaxy mass for high-density relative to low-density environments.

1 Introduction

The role of environment in shaping the properties of ETGs – the most massive galaxies at $z \sim 0$ – is still an open question of galaxy formation and evolution. Although galaxy colors do not exhibit any significant dependence on environment

F.L. Barbera (✉)
INAF-OAC, Salita Moiariello 16, 80131, Napoli, Italy
e-mail: labarber@na.astro.it

P.A.A. Lopes
Observatório do Valongo/UFRJ, Rio de Janeiro, Brazil
e-mail: paal05@gmail.com

R.R. de Carvalho
Instituto Nacional de Pesquisas Espaciais/MCT, S. J. dos Campos, Brazil

I. Ferreras and A. Pasquali (eds.) *Environment and the Formation of Galaxies: 30 years later*, Astrophysics and Space Science Proceedings,
DOI 10.1007/978-3-642-20285-8_15,

at fixed luminosity (e.g. [11]), many spectroscopic studies have reported younger luminosity-weighted ages for field relative to cluster ETGs, by ~1–2 Gyr (e.g. [3,19]). This finding has been recently questioned by [20], who find that indeed only a fraction of *rejuvenated* ETGs (those showing signs of an ongoing star formation) changes with the environment. An even more puzzling issue is the difference in metal content among different environments: field ETGs are more metal-rich [19], as metal-rich as [3], or even more metal-poor than their cluster counterparts [9]. The Fundamental Plane (FP) relation of ETGs [7,8], i.e. the scaling law involving radius, velocity dispersion, and surface brightness, is a powerful tool to measure their M/L ratio. One main feature of the FP is the observed "tilt" of its slopes from the expectations of the virial theorem for a homologous family of systems with constant M/L. Contrasting results have been reported about the environmental dependence of the FP. [3] found the FPs at low - and high -density to have consistent slopes but significant offset, corresponding to a difference of ~1 Gyr in the formation epoch of field and cluster ETGs. Other studies have found a consistent redshift evolution of the M/L ratio of ETGs among different environments, implying the same formation epoch of field and cluster ETGs (e.g. [6, 21]). In this contribution, we focus on the dependence of the FP on local galaxy density, at both optical and NIR wavebands. The analysis is part of the Spheroid's Panchromatic Investigation in Different Environmental Regions (SPIDER; see [14]).

2 The Sample

The SPIDER ETGs are selected from SDSS-DR6 (see [14]) by the *eClass* and *fracDev*$_r$ attributes, as in [2]. Out of 39,993 galaxies with $0.05 \leq z \leq 0.095$ and $M_r < -20$ (optical sample), 5,080 (optical+NIR sample) have photometry in the *YJHK* wavebands from UKIDSS-LAS [17]. In all wavebands, effective parameters, i.e. the effective radius, R_e, and the mean surface brightness within that radius, $<\mu>_e$, have been homogeneously measured with 2DPHOT [13]. Central velocity dispersions, σ_0, are retrieved from SDSS, and corrected to an aperture of radius $R_e/8$ [12].

The environment where ETGs reside is characterized by updating the FoF group catalogue of [1] on SDSS-DR7. We establish group galaxy membership by the "shifting gapper" technique as in [18]. Projected local density, Σ_N, is measured for each galaxy with respect to other group members, considering the distance to the Nth nearest neighbor, where N scales with group richness. This allows us to split the optical (optical+NIR) sample of ETGs into two subsamples of 16,717 (2,187) group and 11,824 (1,359) field galaxies.

3 The Fundamental Plane as a Function of Environment

We write the FP relation as:

$$\log R_e = a \log \sigma_0 + b <\mu>_e + c, \quad (1)$$

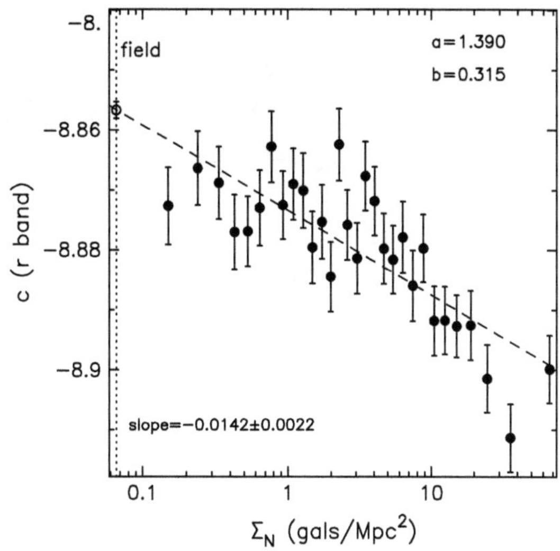

Fig. 1 Offset "*c*" of the FP as a function of local galaxy density, Σ_N. *Filled symbols* represent the binning of the group sample of ETGs with respect to Σ_N. *Error bars* denote 1 σ uncertainties. The *empty circle* corresponds to the field sample. The "*c*" has been computed by fixing "*a*" and "*b*" in all bins to the same values of 1.39 and 0.315, respectively (see [15]). The *dashed line* is the least-squares fit to the data

where "*a*" and "*b*" are the slopes, and "*c*" is the offset (intercept). The values of "*a*", "*b*", and "*c*", are estimated by an orthogonal robust fitting procedure, accounting for different selection effects (see [15, 16] for details). Figure 1 plots the "*c*" for the optical sample of ETGs, in different bins of local galaxy density, Σ_N, each bin including the same number of 557 galaxies. The FP intercept smoothly decreases from low (i.e. field) to high density regions (cluster cores), consistent with [3]. Interpreting the FP as the virial theorem in action, under the assumption of homology, the "*c*" is inversely proportional to the average mass-to-light ratio (M/L) of ETGs. Since we do not detect any environmental variation of the typical dynamical mass of the sample (see [16] for details), we interpret the trend in Fig. 1 as a smooth decrease of galaxy luminosity, at fixed mass, with density. The luminosity variation per density decade is given by the slope, b_{Σ_N}, of the linear fit in Fig. 1. Figure 2 plots the b_{Σ_N} as a function of the waveband (*grizYJHK*). Remarkably, b_{Σ_N} does not change from g through K. Different curves show stellar population models of b_{Σ_N}. Each model consists of a pair of SSPs from [4], with given difference in age ($\delta_t = \delta \log t / \delta \log \Sigma_N$) and metallicity ($\delta_Z = \delta \log Z / \delta \log \Sigma_N$). The best-fitting model (solid curve in the figure) has $\delta_Z = -0.043 \pm 0.023$ and $\delta_t = 0.048 \pm 0.006$, i.e. ETGs at high density are older, by $\sim$11% per density decade, and less metal-rich than those at low density. The δ_Z is consistent with zero at $\sim 2\sigma$, implying that a pure age model (dashed curve) might also be able by itself to explain the data.

Figure 3 plots the slopes of the FP as a function of waveband, for group, and field ETGs. The "*a*" increases from g through K, for both samples, by $\sim$15–19%. The behavior of b depends on environment. The field ETGs exhibit a small, but significant variation ($\sim$2.5%) with waveband, while for group ETGs the "*b*" is constant. In other words, field ETGs exhibit a stronger dependence of FP slopes on waveband, as both "*a*" and "*b*" increase from g through K. Notice also that group

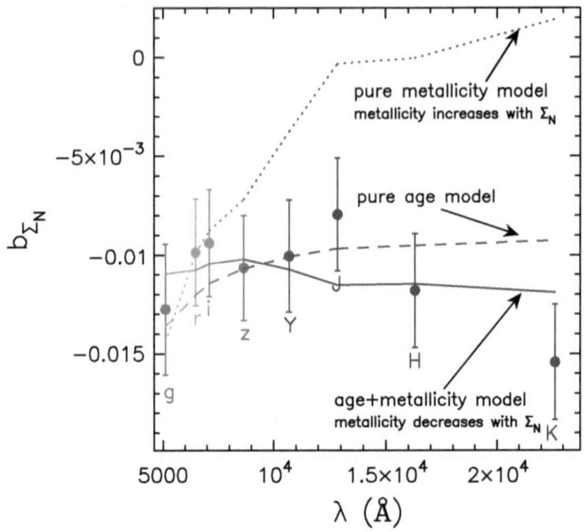

Fig. 2 Difference of FP intercept per decade in local density, $\delta c/\delta \log \Sigma_N$, as a function of the effective wavelength of each filter (*grizYJHK*). The *solid*, *dashed*, and *dotted curves* are age+metallicity ($A + M$), pure age (A), and pure metallicity (M) stellar population models, where low-density galaxies are younger and more metal rich ($A + M$), younger (A), and less metal rich (M), than their high-density counterparts. Notice how model $A + M$ reproduces the observed trend

ETGs have smaller "*a*" relative to field galaxies, at all wavebands. On the other hand, "*b*" is larger in the optical for group ETGs, with the difference diminishing in the NIR.

4 Conclusions

We find that the FP offset smoothly decreases from low- to high-density environments, in the same way from g through K. Assuming that the age and metallicity of galaxy stellar populations are the main drivers of this variation, this implies that low-density ETGs have to be younger and more metal-rich (at $2\,\sigma$) than their cluster counterparts. The former finding is fully consistent with the expectations of semi-analytical models of galaxy formation (e.g. [5]), while the latter is not. Assuming that the K-band tilt of the FP results from a variation of dark matter fraction with mass (see [15] and references therein), the fact that group ETGs have lower "*a*" (i.e. larger tilt) *wrt* field galaxies at all wavebands, would mean that group ETGs have a steeper variation of dark matter fraction with mass. The stronger dependence of FP slopes on waveband for field ETGs would also imply that the correlation of stellar population properties with mass is shallower for high-density relative to low-density environments.

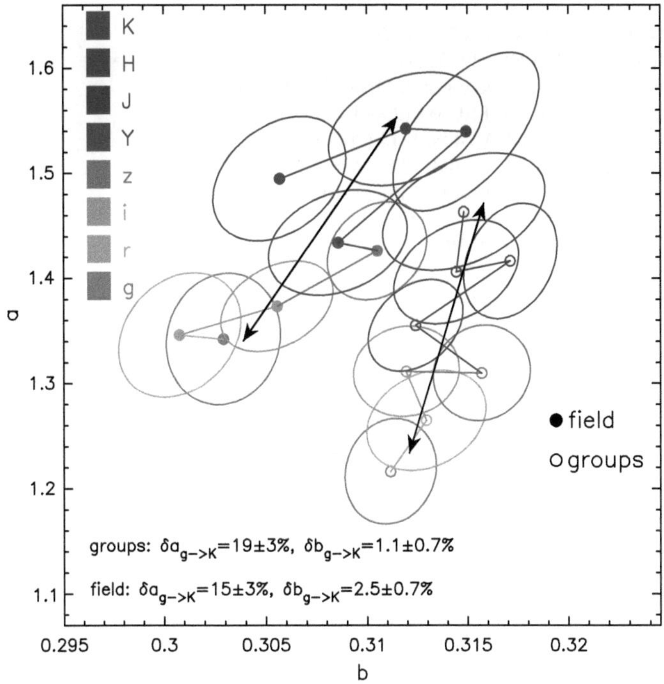

Fig. 3 Slopes "*a*" and "*b*" of the FP for galaxies residing in groups (*empty circles*) and in the field (*filled circles*) as a function of waveband. Different *grey* levels correspond to different wavebands (see *upper-left*), from *g* through *K*. *Ellipses* denote 1 σ confidence contours on "*a*" and "*b*". The *black arrows* plot the variation of the FP slopes from *g* through *K*. The relative variations, $\delta a_{g->K}$ and $\delta b_{g->K}$, are reported in the *lower-left* of the plot

Acknowledgements We acknowledge the use of SDSS data.[1] We have used data from the 4th data release of the UKIDSS survey, which is described in detail in [22] (see also [10]).

References

1. A.A. Berlind et al., ApJS **167**, 1 (2006)
2. M. Bernardi, R.K. Sheth, J. Annis, AJ **125**, 1849 (2003)
3. M. Bernardi et al., AJ **131**, 1288 (2006)
4. G. Bruzual, S. Charlot, MNRAS **344**, 1000 (2003)
5. G. de Lucia et al., MNRAS **366**, 499 (2006)
6. S. di Serego Alighieri, B. Lanzoni, I. Jörgensen, ApJ **647**, 99 (2006)
7. S.G. Djorgovski, M. Davis, ApJ **313**, 59 (1987)
8. A. Dressler et al., ApJ **313**, 42 (1987)

[1] http://www.sdss.org/collaboration/credits.html.

9. A. Gallazzi et al., MNRAS **370**, 1106 (2006)
10. P.C. Hewett et al., MNRAS **367**, 454 (2006)
11. D.W. Hogg et al., ApJ **601**, 29 (2004)
12. I. Jørgensen, M. Franx, P. Kjærgaard, MNRAS **276**, 1341 (1995)
13. F. La Barbera et al., PASP **120**, 681 (2008)
14. F. La Barbera et al., MNRAS **408**, 1313 (2010)
15. F. La Barbera et al., MNRAS **408**, 1335 (2010)
16. F. La Barbera et al., MNRAS **408**, 1361 (2010)
17. A. Lawrence et al., MNRAS **379**, 1599 (2007)
18. P.A.A. Lopes et al., MNRAS **392**, 135 (2009)
19. D. Thomas et al., ApJ **621**, 673 (2005)
20. D. Thomas et al., MNRAS **404**, 1775 (2010)
21. P.G. van Dokkum, R.P. van der Marel, ApJ **655**, 30 (2006)
22. S.J. Warren et al., MNRAS **375**, 213 (2007)

On Galaxy Mass–Radius Relationship

D. Bindoni, L. Secco, E. Contini, and R. Caimmi

Abstract In the Clausius' virial maximum theory (TCV) [Secco and Bindoni, NewA 14, 567 (2009)] to explain the galaxy Fundamental Plane (FP) a natural explanation follows about the observed relationship between stellar mass and effective radius, $M_* - r_e$, for early type galaxies (ETGs). The key of this correlation lies in the deep link which has to exist between cosmology and the existence of the FP. The general strategy consists in using the two-component tensor virial theorem to describe the virial configuration of the baryonic component of mass $M_B \simeq M_*$ embedded in a dark matter (DM) halo of mass M_D at the end of relaxation phase. In a ΛCDM flat cosmology, starting from variance at equivalence epoch, we derive some preliminary theoretical relationships, $M_* - r_e$, which are functions of mass ratio $m = M_D/M_B$. They appear to be in agreement with the trends extracted from the data of galaxy sample used by [Tortora et al., MNRAS 396, 1132 (2009)].

1 The Model and the Cosmological Framework

In the TCV expectation [2], a special scale length is induced by the DM halo on the baryonic gravitational field. This dimension, among the infinite possible ones corresponding to virial equilibrium, maximizes the Clausius' virial energy distributing it in about equal amounts between baryons and DM. Consequently, the visible virialized matter has a special dimension, the tidal radius a_t, strictly related to the DM virial radius, a_D. We assume for the bright component a spatial matter density profile obtained from the empirical surface light density law proposed by King (1962). A cored power-law density profile is assumed for the DM distribution. The framework is a ΛCDM model of parameters: $\sigma_8 = 0.9 \pm 0.1$; $h = 0.72 \pm 0.05$;

D. Bindoni · L. Secco (✉) · E. Contini · R. Caimmi
Department of Astronomy, vic. Osservatorio 3, 35122 Padova – I, Italy
e-mail: luigi.secco@unipd.it

I. Ferreras and A. Pasquali (eds.) *Environment and the Formation of Galaxies: 30 years later*, Astrophysics and Space Science Proceedings,
DOI 10.1007/978-3-642-20285-8_16,

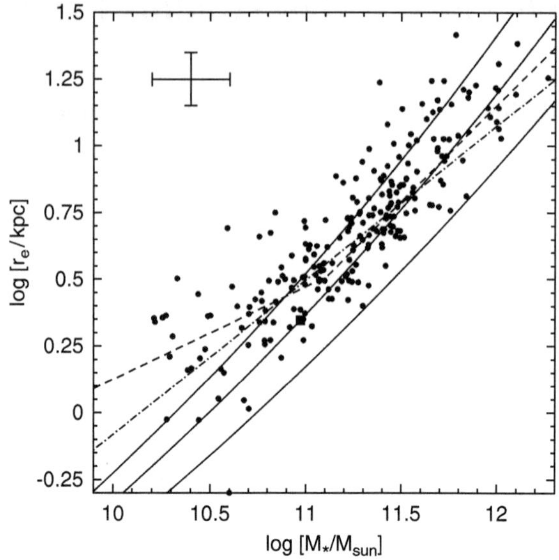

Fig. 1 Masses and radii of local ETGs from [4]. The *dashed line* is the fit to the data, whose slope changes from the low- to the high-mass regime. The *point-dashed line* shows the mean mass-size relation. The *solid lines* are the theoretical *curves* given by (1), when the mass ratio is $m = 1, 4, 10$ (*bottom-up*) and $d = 0.5$. The *black square* represents the calibration galaxy of the sample for $m_0 = 4$

$\Omega_m h^2 = 0.14 \pm 0.02$; $\Omega_b h^2 = 0.024 \pm 0.001$ and primordial spectral index $n_s = 0.99 \pm 0.04$ consistent with the Harrison–Zeldovich' scale invariant one ($n_s = 1$) derived from WMAP precision data [3]. In this self-similar model, an effective final spectrum index, n_e, may be obtained from the local slope of the mass-variance σ_{M_D} at a given M_D value. At the halo virialization the following key scaling law holds: $r_e \sim m^r M_B^R$; $r = [(3-d) - \gamma']/[\gamma'(3-d)]$; and $R = 1/\gamma'$, where d is the density profile of DM and γ' is related to the cosmological local slope of the mass variance.

Theoretical fits of data: We refer to the mass-size relation extracted by the data set of [1] used in a dynamical study of local ETGs [4]. Our theoretical mass-size relationship differentiates both the mass ratio m contribution and the trend of the slope in order to produce a set of continue curves on the data plane ($\log M_*$, $\log r_e$). That improves the resolution of the relationship and allows a comparison with the trends for individual galaxies of the sample (Fig. 1).

2 Conclusion

By means of the TCV we have derived some preliminary mass-size relationships for ETGs in a ΛCDM cosmological scenario that are in agreement with the mean one found by fitting the binned averages data of [1] by [4]. We emphasize that their slope is really a mean one because the theoretical slope changes as soon as the mass changes. Moreover the linear fit of [4] is a mean one also because it goes across the theoretical lines, each of them being characterized by dark/bright mass ratio, m. The existence of individual galaxies with different values of this ratio actually produces a real scatter. Not taking this intrinsic scatter into account makes it is impossible to determine the mean theoretical slope in the sample, either in the lower mass regime or in the higher one in order, and to compare the mean theoretical slope with those derived by [4].

References

1. P. Prugniel, F. Simien, A&A **309**, 749 (1996)
2. L. Secco, D. Bindoni, NewA **14**, 567 (2009)
3. D.N. Spergel et al., ApJS **170**, 377 (2007)
4. C. Tortora et al., MNRAS **396**, 1132 (2009)
5. I. King, AJ **67**, 274 (1962)

And the Winner Is: Galaxy Mass

D. Thomas

Abstract Environment is known to affect the formation and evolution of galaxies considerably, and this is best visible through the well-known morphology–density relationship. We study the effect of environment on the evolution of early-type galaxies for a sample of 3,360 galaxies morphologically selected by visual inspection from the SDSS in the redshift range $0.05 \leq z \leq 0.06$, and analyse luminosity-weighted age, metallicity, and α/Fe ratio as function of environment and galaxy mass. We find that on average 10% of early-type galaxies are rejuvenated through minor recent star formation. This fraction increases with both decreasing galaxy mass and decreasing environmental density. However, the bulk of the population obeys a well-defined scaling of age, metallicity, and α/Fe ratio with galaxy mass that is independent of environment. Our results contribute to the growing evidence in the recent literature that galaxy mass is the major driver of galaxy formation. Even the morphology–density relationship may actually be mass-driven, as the consequence of an environment dependent characteristic galaxy mass coupled with the fact that late-type galaxy morphologies are more prevalent in low-mass galaxies.

1 Introduction

Environment is known to be a major driver in the formation and evolution of galaxies. Its influence is best visible through the well-known morphology–density relationship, according to which early-type galaxies and morphologically undisturbed galaxies are preferentially found in high density environments and vice versa [3, 9].

D. Thomas (✉)
Institute of Cosmology and Gravitation, University of Portsmouth, Dennis Sciama Building, Burnaby Road, Portsmouth PO1 3FX, UK
e-mail: daniel.thomas@port.ac.uk

I. Ferreras and A. Pasquali (eds.) *Environment and the Formation of Galaxies: 30 years later*, Astrophysics and Space Science Proceedings,
DOI 10.1007/978-3-642-20285-8_17, © Springer-Verlag Berlin Heidelberg 2011

In contrast, it is less clear whether the environment is equally important at a given galaxy morphology. There is still major controversy about whether the formation and evolution of the most massive and morphologically most regular galaxies in the universe, i.e. early-type galaxies, are affected by environmental densities. We analyse the stellar population parameters luminosity-weighted age, metallicity, and α/Fe element ratio of 3,360 early-type galaxies drawn from the Sloan Digital Sky Survey [18] in a narrow redshift range ($0.05 \leq z \leq 0.06$) [16]. The SDSS provides the opportunity to explore huge homogeneous samples of early-type galaxies in the nearby universe, comprising several ten thousands of objects, so that a statistically meaningful investigation of the stellar population parameters of galaxies and their dependence on environment can be attempted. This work is published in [16], and we refer the reader to this paper for more details. The galaxy catalogue produced in this study can be found at www.icg.port.ac.uk/~thomasd.

2 Data

The sample utilised here is part of a project called MOSES: **MO**rphologically **S**elected **E**arly-types in **S**DSS. We have collected a magnitude limited sample of 48,023 galaxies in the redshift range $0.05 \leq z \leq 0.1$ with apparent r-band magnitude brighter than 16.8 from the SDSS Data Release 4. The most radical difference with respect to other galaxy samples constructed from SDSS is our choice of purely morphological selection of galaxy type through visual inspection [12]. Building on the success of this strategy, our approach has been extended with the Citizen Science Galaxy Zoo project to enable visual classification of even larger samples [6].

For the estimate of the environmental density we calculate the local number density of objects brighter than a certain absolute magnitude in a 6 Mpc sphere around the object of interest with a Gaussian weight function centred on the object to avoid contamination from neighbouring structures. We compute stellar velocity dispersion and measure emission line fluxes, which we use to produce an emission line free galaxy spectrum with the publicly available codes ppxf and GANDALF [2, 11]. We measure Lick absorption-line indices from the spectra, and derive luminosity-weighted ages, metallicities, and α/Fe ratios using the TMB stellar population models [14] by means of a minimised χ^2 technique.

We applied further selection criteria to the 16,502 early-type galaxies in our catalogue. The redshift range sampled ($0.05 \leq z \leq 0.1$) is small and corresponds to a time window of only ~600 Myr. Still, this age difference produces significant selection effects. We therefore took a conservative approach and decided to focus on the narrow redshift range $0.05 \leq z \leq 0.06$, providing an acceptable coverage in velocity dispersion down to $\log \sigma \text{ km s}^{-1} \sim 1.9$. This results in a total of 3,360 early-type galaxies.

3 Results

The characteristic mass of a galaxy population increases with environmental density [1], indicating that lower density environments host less massive galaxies. This mass bias has to be eliminated for a meaningful study of the influence of environment. We therefore investigate the environmental dependence of the correlations of the stellar population properties with velocity dispersion σ and galaxy mass.

Figure 1 shows contour plots for the relationship between stellar velocity dispersion and luminosity-weighted age for various environmental densities as indicated by the inset histograms. The distribution of ages is bimodal with a major peak at old ages and a secondary peak at young ages around ∼2.5 Gyr in analogy to 'red sequence' and 'blue cloud' identified in galaxy populations. We call the objects below the dotted line 'rejuvenated', and the fraction of objects in this category 'rejuvenation fraction'. The latter is about 10% as indicated by the label in the figure.

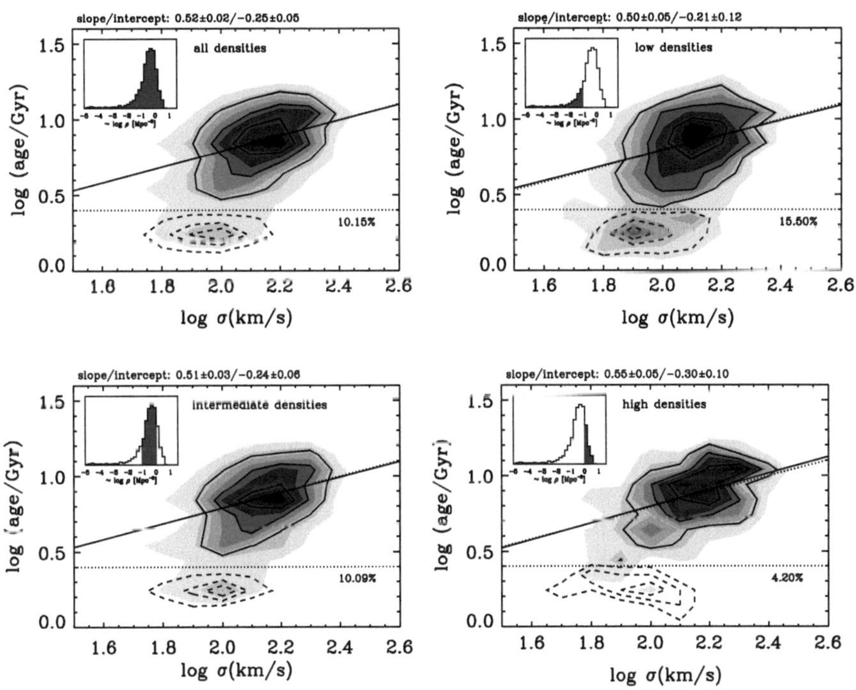

Fig. 1 Contour plots of the relationship between stellar velocity dispersion and luminosity-weighted age for various environmental densities as indicated by the inset histograms. The horizontal *dotted line* separates an old red sequence population (*solid contours*) from rejuvenated objects in the *blue* cloud with light-averaged ages smaller than 2.5 Gyr (*dashed contours*). The *solid line* is a linear fit to the *red* sequence population, the parameters of the fit are given at the *top* of each panel. The *dotted line* is the fit for all environmental densities (parameters from *top left-hand panel*). The age-σ relationship for the *red* sequence population is independent of environment, while the rejuvenation fraction increases with decreasing density

The fits to the red sequence population are repeated in every environment bin. It can be clearly seen from the plots that the resulting fit parameters are consistent with no variation as a function of environment within their 1-σ error bars. We conclude there is no considerable change as a function of environmental density, hence the age–σ relationship is independent of environment. The same is true for metallicity and α/Fe ratio [16].

Very different is the behaviour of the blue cloud (rejuvenated) population. In this case, the environment plays a role. The rejuvenation fraction increases with decreasing environmental density. Hence early-type galaxies in lower density environments are not generally younger (at a given mass), but the fraction of rejuvenated galaxies is higher. Note, however, that again mass is the major driver for the fraction of rejuvenated galaxies as shown in [16].

This environment-dependent rejuvenation process is most prevalent in low-mass galaxies, and entirely absent in the most massive objects. Most importantly, rejuvenation involves only minor star formation, hence is negligible in terms of the overall mass budget, and occurs at late epochs. As a consequence, most of the galaxy's stellar populations form in an environment independent mode with galaxy mass as the major driver for star formation and quenching processes.

4 Discussion

In [16] we investigate the effect of environmental density on the formation epochs of early-type galaxies. The major conclusion is that it is galaxy mass, not environment, that shapes the stellar population properties of early-type galaxies. Pre-SDSS studies of the stellar population parameters in early-type galaxies based on relatively small, local samples consistently found younger average ages in low density environments [5, 15]. The present sample based on SDSS is significantly larger and therefore less biased towards the local volume, which may explain the discrepancy with previous work. The increase in sample size further implies a higher statistical significance of the results. In fact most studies in the recent literature that are based on large samples from data bases such as the SDSS come to similar conclusions [4, 8, 10, 17]: galaxy mass dominates over galaxy environment. In particular, environment seems to be important mostly for low-mass galaxies and only at late epochs [7, 13].

Based on a study of SDSS data, [17] show that the colours and concentrations of satellite galaxies are determined by their stellar mass. In particular, at fixed stellar mass, they find the average colours and concentrations of satellite galaxies to be independent of either halo mass or halo-centric radius [17]. The only fingerprint from the environment is that satellites appear to be redder, older and metal-richer than centrals of the same stellar mass, a difference that increases with decreasing galaxy mass [7]. However, they find that the nature of the transformation and quenching process experienced by a galaxy when it falls into a bigger halo is independent of the size of this halo, hence independent of environment [17].

[9] use SDSS data to separate two distinct processes of 'mass-quenching' and 'environment-quenching' that dominate the evolution of galaxies. They find that the quenching of galaxies around and above M^* must follow a rate that is statistically proportional to their star-formation rates, and the latter tend to scale with mass rather than with environment. [9] conclude that the environment acts through a 'once-only' process as the environment of a given galaxy changes, while the mass-quenching process must be continuously operating and be governed by a probabilistic transformation rate. Again, this result implies that environment only plays a secondary role in shaping the stellar populations of galaxies, and galaxy mass is the major driver of galaxy formation.

5 Conclusions

The role and significance of environment for galaxy formation has been discussed highly controversially at the JENAM 2010 Symposium 2 'Environment and the Formation of Galaxies: 30 years later'. It is exciting and intriguing that recent work is about to lead to a converging picture, despite some controversy, in which environment loses out over galaxy mass, 30 years after the seminal work of Alan Dressler on the morphology–density relationship [3]. As it turned out at this meeting, however, Alan Dressler has never regarded the morphology–density relation necessarily as environment-driven. It could actually be mass-driven, being the consequence of an environment dependent galaxy mass function (the characteristic mass of a galaxy population increases with environmental density) coupled with the fact that late type galaxy morphologies are more prevalent in low-mass galaxies. Galaxies are individualists, and, as Alan Dressler concluded at the meeting, 'morphology is destiny'.

References

1. I.K. Baldry et al., MNRAS **373**, 469 (2006)
2. M. Cappellari, E. Emsellem, PASP **116**, 138 (2004)
3. A. Dressler, ApJ **236**, 351 (1980)
4. R. Grützbauch et al., MNRAS **411**, 929 (2011). arXiv:1009.3189
5. H. Kuntschner et al., MNRAS **337**, 172 (2002)
6. C.J. Lintott et al., MNRAS **389**, 1179 (2008)
7. A. Pasquali et al., MNRAS **407**, 937 (2010)
8. Y.J. Peng, S. Lilly, ApJ **721**, 193 (2010)
9. M. Postman, M.J. Geller, ApJ **281**, 95 (1984)
10. B. Rogers et al., MNRAS **405**, 329 (2010)
11. M. Sarzi et al., MNRAS **366**, 1151 (2006)
12. K. Schawinski et al., MNRAS **382**, 1415 (2007)
13. L.A.M. Tasca et al., A&A **503**, 379 (2009)

14. D. Thomas, C. Maraston, R. Bender, MNRAS **339**, 897 (2003)
15. D. Thomas et al., ApJ **621**, 673 (2005)
16. D. Thomas et al., MNRAS **404**, 1775 (2010)
17. F.C. van den Bosch et al. (2008, preprint). arXiv:0805.0002
18. D.G. York et al., AJ **120**, 1579 (2000)

Ages of Globular Cluster Systems and the Relation to Galaxy Morphology

A.L. Chies-Santos, S.S. Larsen, H. Kuntschner, P. Anders, E.M. Wehner, J. Strader, J.P. Brodie, and J.F.C. Santos Jr.

Abstract Some photometric studies of extragalactic globular cluster (GC) systems using the optical and near-infrared colour combination have suggested the presence of a large fraction of intermediate-age (2–8 Gyrs) GCs. We investigate the age distributions of GC systems in 14 E/S0 galaxies. We carry out a differential comparison of the $(g - z)$ *vs.* $(g - K)$ two-colour diagrams for GC systems in the different galaxies in order to see whether there are indications of age differences. We also compare the different GC systems with a few simple stellar population models. No significant difference is detected in the mean ages of GCs among elliptical galaxies. S0 galaxies on the other hand, show evidence for younger GCs. Surprisingly, this appears to be driven by the more metal-poor clusters. The age distribution of GCs in NGC 4365 seems to be similar to that of other large ellipticals (e.g. NGC 4486, NGC 4649). Padova SSPs with recently released isochrones for old ages (14 Gyrs) show less of an offset with respect to the photometry than previously published models. We suggest that E type galaxies assembled most of their GCs in a shorter and earlier period than S0 type galaxies. The latter galaxy type seems to have a more extended period of GC formation/assembly.

A.L. Chies-Santos (✉) · S.S. Larsen · P. Anders · E.M. Wehner
Astronomical Institute Utrecht, Princetonplein 5, 3584, Utrecht, The Netherlands
e-mail: A.L.Chies@uu.nl

H. Kuntschner
Space Telescope European Coordinating Facility, European Southern Observatory, Garching, Germany

J. Strader
Harvard-Smithsonian Center for Astrophysics, Cambridge, MA 02138, USA

J.P. Brodie
UCO/Lick Observatory, University of California, Santa Cruz, CA 95064, USA

J.F.C. Santos Jr.
Departamento de Física, ICEx, Universidade Federal de Minas Gerais, Av. Antônio Carlos 6627, Belo Horizonte 31270-901, MG, Brazil

I. Ferreras and A. Pasquali (eds.) *Environment and the Formation of Galaxies: 30 years later*, Astrophysics and Space Science Proceedings,
DOI 10.1007/978-3-642-20285-8_18,

1 Deriving Ages of GC Systems Differentially

It is well documented in the literature that a direct comparison of integrated GC colours with simple stellar population (SSP) models tends to yield ages that are significantly younger than a Hubble time (e.g. [3]).

We derive optical and near-infrared colours for GC systems of 14 early-type galaxies in [1]. In order to avoid SSP models on the derivation of GC system ages, we turn to a purely differential comparison. The median of the age distribution of the GC system of NGC 4486 is reported to be 13 Gyrs with a dispersion of 2 Gyrs, through spectroscopy ([2]). The GCs of NGC 4649 are distributed in the $(g - K)$ *vs.* $(g - z)$ plane in a similar way to those of NGC 4486. Being these two GC systems supposedly old, and the ones with more clusters in our sample (167 and 301 respectively) we take them as the fiducial old systems. We define best fit lines in the $(g - K)$ *vs.* $(g - z)$ plane for all the clusters together and by separating in blue and red clusters. The upper left panel of Fig. 1 shows a cartoon illustrating a parameter which we define as δ for a cluster in the $(g - K)$ vs. $(g - z)$ plane. The best fit line for the combined GC system of NGC 4486 and NGC 4649 is also shown in this figure. The location of younger and older clusters relative to this line and the location of the metal-rich and metal-poor clusters is also indicated. The values of δ

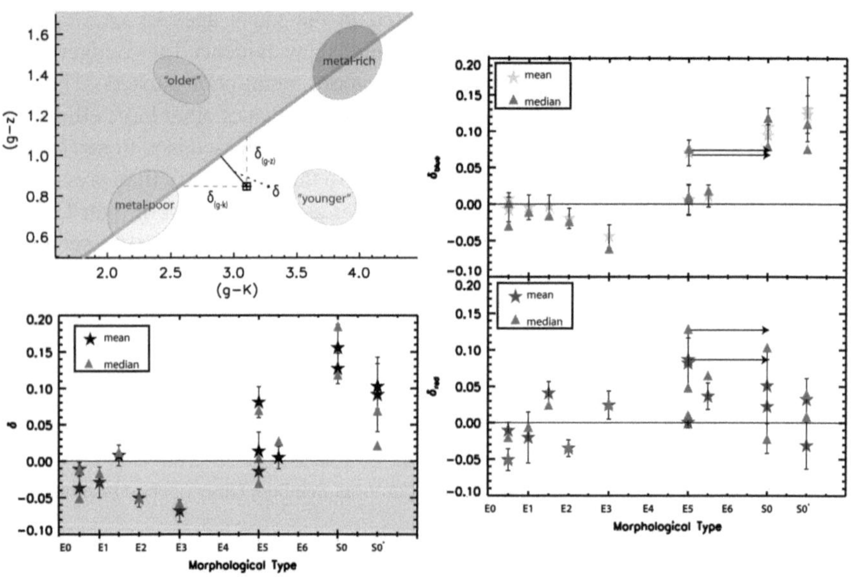

Fig. 1 *Upper left panel*: A cartoon illustrating the definition of δ, an age proxy in the $(g - k)$ vs. $(g - z)$ for one cluster and the location of the younger and older objects relative to the *solid line*. The location of metal-poor and metal-rich objects is also shown. *Bottom left panel*: The mean and median δ values as a function of morphological type for each galaxy. *Upper right panel:* The mean and median $\delta_{\rm blue}$ values as a function of galaxy morphology. *Bottom right panel:* The mean and median $\delta_{\rm red}$ values as a function of galaxy morphology

are age tracers of GCs in this plane. An object falling to the right of the solid line will have a positive δ value. A positive δ value means that this object has an age that is younger than the age that the solid line traces ($\delta = 0$). In this sense a GC system whose mean difference is further away to the right of the $\delta = 0$ value in the distributions is the system with the best chances of hosting younger GCs.

In the bottom left panel of Fig. 1 the mean (and median) of δ is plotted against the morphological type of the galaxy. It is readily seen that the mean of δ correlates with galaxy morphology, with S0 galaxies having larger δ values compared to E's. Surprisingly, this relation is stronger when only the blue (metal-poor) population is considered (upper right panel). The trend for the red (metal-rich) clusters is less clear, as can be seen in the bottom right panel of Fig. 1.

References

1. A.L. Chies-Santos et al., A&A **525**, 19 (Paper I) (2011)
2. J.G. Cohen, J.P. Blakeslee, A. Ryzhov, ApJ **496**, 808 (1998)
3. S.S. Larsen, J.P. Brodie, J. Strader, A&A **443**, 413 (2005)

AGN Feedback and Quenching of Star Formation: A Multiwavelength Approach with the EURO-VO

B. Coelho, C. Lobo, and S. Antón

Abstract We selected bulgeless red sequence galaxies in the SDSS (DR7) [Abazajian et al., ApJS 182, 543 (2009)] using data from NYU-VAGC [Blanton et al., AJ 129, 2562 (2005)]. Using EURO-VO tools we obtained multiwavelength data, we built spectral energy distributions, and undertook a thorough analysis to ascertain: the frequency of AGN among these galaxies, the degree of star formation and intrinsic extinction. We aim at verifying whether there are bulgeless quenched galaxies hosting SMBHs in order to test the AGN feedback paradigm.

1 Sample and Data

Using the NYU-VAGC we selected SDSS DR7 [1,2] galaxies with: redshift between 0.02 and 0.06, stellar mass, M_*, larger than 10^{10} M_{sun}, bulge component with Sérsic index n<1.5, inclination lower than 60° to minimize dust extinction effects and colour index $g - r > 0.57 + 0.0575 * log(M_*/10^8 M_{sun})$ (Fig. 1).

VO-tools revealed to be extremely useful: e.g. TOPCAT and STILTS to deal with data tables, and VO-Desktop to obtain multiwavelength fluxes (Fig. 2).

B. Coelho (✉) · C. Lobo
Centro de Astrofisica da Universidade do Porto, Rua das estrelas 4150-762 Porto, Portugal
and
Faculdade de Ciencias, Universidade do Porto, Rua do Campo Alegre 4169-007 Porto, Portugal
e-mail: up000308008@alunos.fc.up.pt

S. Antón
CICGE, Faculdade de Ciencias, Universidade do Porto, Rua do Campo Alegre 4169-007 Porto, Portugal
and
SIM, Faculdade de Ciencias, Universidade de Lisboa, Campo Grande 1649-016 Lisboa, Portugal

I. Ferreras and A. Pasquali (eds.) *Environment and the Formation of Galaxies: 30 years later*, Astrophysics and Space Science Proceedings,
DOI 10.1007/978-3-642-20285-8_19, © Springer-Verlag Berlin Heidelberg 2011

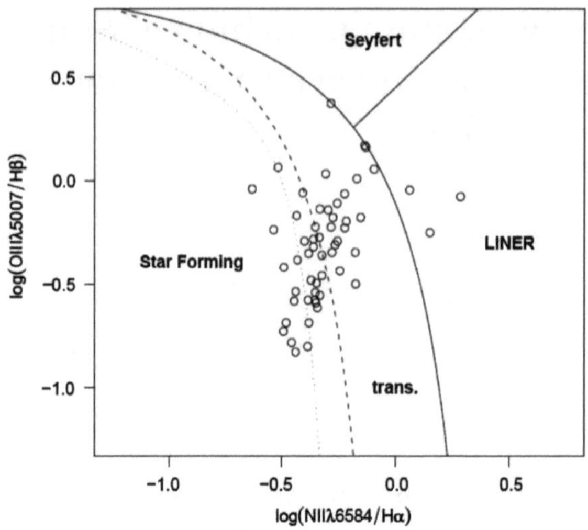

Fig. 1 *[OIII]*$5007/H_\beta$ vs. *[NII]*$6584/H_\alpha$ diagram. The *solid curve* [4] separates pure AGN hosts from galaxies with SF. *Dashed* [3], *dotted* [6] and *straight line* [5]

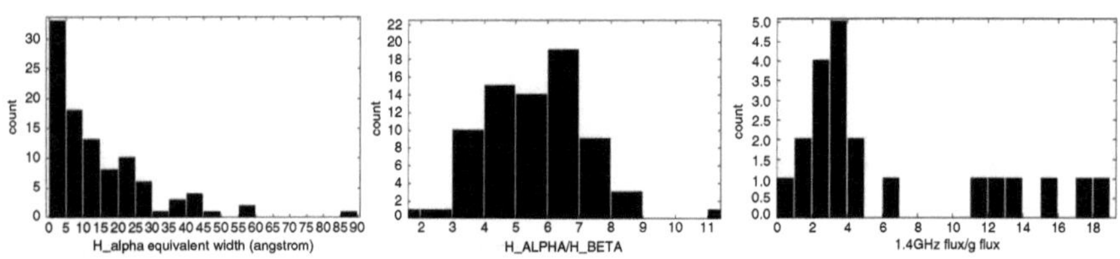

Fig. 2 *Left* – Histogram of H_α equivalent width. *Middle* – H_α/H_β. Despite the inclination cut, some objects are affected by extinction. *Right* – 1.4 GHz (FIRST)/g(SDSS) flux ratio. The radio sources of our sample are mainly radio-quiet and some are "radio-intermediate" sources

Emission line fluxes were obtained from the MPA-JHU DR7 measurements (http://www.mpa-garching.mpg.de/SDSS/DR7/).

2 Data Analysis

We used emission-line diagnostics to check for the existence of AGNs and check on the presence of dust and star formation (SF) in the 113 objects of the sample (Fig. 1).

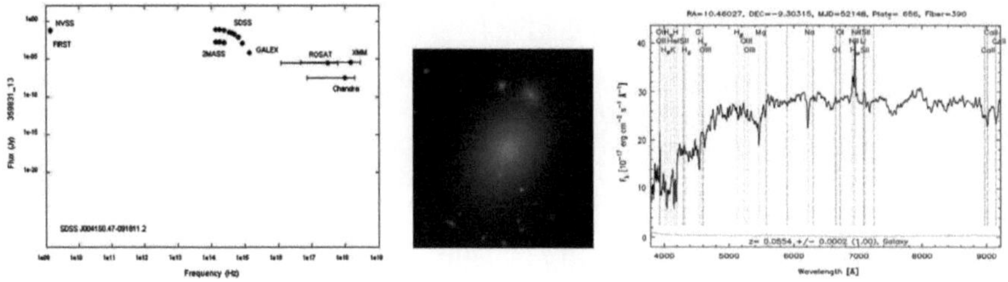

Fig. 3 SDSS J004150.47-091811.2: *Left* – spectral energy distribution. *Middle* – SDSS g, r and i composite image. *Right* – SDSS spectrum. This source is a bulgeless galaxy with a LINER spectrum. Its location in the BPT diagram suggests that no star formation is taking place

3 Conclusions

We have found "red sequence" galaxies with star formation and dusty regions (Fig. 2), that were previously assumed to be quenched galaxies.

We have also found one bulgeless galaxy (Fig. 3) that: (1) harbors an AGN (2) based on BPT diagram, no star formation is taking place. It is our best candidate for a bulgeless quenched galaxy hosting a SMBH.

References

1. K.N. Abazajian et al., ApJS **182**, 543 (2009)
2. M.R. Blanton et al., AJ **129**, 2562 (2005)
3. G. Kauffmann et al., MNRAS **346**, 1055 (2003)
4. L. Kewley et al., ApJ **556**, 121 (2001)
5. K. Schawinski et al., MNRAS **382**, 1415 (2007)
6. G. Stasińska et al., MNRAS **371**, 972 (2006)

Does Environment Affect the Star Formation Histories of Early-Type Galaxies?

I. Ferreras, A. Pasquali, and B. Rogers

Abstract Differences in the stellar populations of galaxies can be used to quantify the effect of environment on the star formation history. We target a sample of early-type galaxies from the Sloan Digital Sky Survey in two different environmental regimes: close pairs and a general sample where environment is measured by the mass of their host dark matter halo. We apply a blind source separation technique based on principal component analysis, from which we define two parameters that correlate, respectively, with the average stellar age (η) and with the presence of recent star formation (ζ) from the spectral energy distribution of the galaxy. We find that environment leaves a second order imprint on the spectra, whereas local properties – such as internal velocity dispersion – obey a much stronger correlation with the stellar age distribution.

1 The Environment of Early-Type Galaxies

Within the standard framework of structure formation, galaxies grow in a hierarchical fashion from small structures, progressively merging into more massive systems. Galaxies in regions with a higher over-density will collapse earlier than galaxies in under-dense regions. Hence, we expect a significant dependence of the star formation histories of galaxies with the environment were they form and evolve.

I. Ferreras (✉)
MSSL, University College London, Holmbury St Mary, Dorking, Surrey RH5 6NT, UK
e-mail: ferreras@star.ucl.ac.uk

A. Pasquali
Astronomisches Rechen-Institut, Zentrum für Astronomie der Universität Heidelberg, Mönchhofstr. 12-14, 69120 Heidelberg, Germany
e-mail: pasquali@ari.uni-heidelberg.de

B. Rogers
Department of Physics, King's College London, Strand, London WC2R 2LS, UK

I. Ferreras and A. Pasquali (eds.) *Environment and the Formation of Galaxies: 30 years later*, Astrophysics and Space Science Proceedings,
DOI 10.1007/978-3-642-20285-8_20,

The morphology-density relation [4] whose 30 year anniversary we celebrate in this Symposium is indeed proof of the fact that environmental mechanisms are important in shaping the galaxy populations we see today. This contribution focuses on the star formation histories of a type of galaxies that are especially sensitive tracers of environment.

The dynamical state of elliptical galaxies suggests a formation process driven by galaxy-galaxy interactions. The current interpretation for their formation history involves major mergers, although an observational quantification of the role of mergers and their impact on the underlying stellar populations remains an open question [3, 6]. This morphological type is thus an optimal target to understand the effect of environment on galaxy formation. In order to quantify the effect of environment, we compare large data sets comprising spectral energy distributions of early-type galaxies in different environmental regimes. Differences in their properties are determined by the application of a blind-source separation method whereby the data alone – no modelling – are used to define an observable that discriminates between galaxies based on their spectral information.

The work presented here is based on spectra from the Sloan Digital Sky Survey (SDSS, [12]). By applying Principal Component Analysis (PCA) to the spectra, we find that most of the information locked in the data (over 99% in the sense of variance) resides in the first two principal components. Projecting all spectra on to a rotated version of these two components allowed us to generate two PCA-based parameters: η and ζ. The rotation of components in the parameter space spanned by the eigenvectors of the covariance matrix is often used as a method to go beyond a simple decorrelation of the data, such as independent component analysis, where statistical independence between components is sought (see e.g. [7]). By comparing the rotated components with models of population synthesis for a wide range of star formation histories, we find that the projection of a galaxy spectrum on to the η component correlates with average age for metal-rich populations (typical of the types of galaxies explored here). Furthermore, projections on to the ζ component are very sensitive to the presence of recent star formation, as found when matching against NUV fluxes of early-type galaxies from GALEX (see [5, 8] for details).

Two environmental regimes are studied. In the next section we consider close pairs of interacting early-type galaxies. In Sect. 3 we present the results for a general, volume-limited sample of early-type galaxies out to $z < 0.1$, where environment is parameterised according to the mass of the halo where the galaxy resides, using the groups catalogue from [11].

2 Environment Over Small Scales: Close Pairs

We define a sample of close pairs by choosing galaxies from SDSS with a radial velocity difference below 500 km s^{-1} and a physical separation within 30 kpc along the transversal direction. This sample is then visually classified, selecting only those systems consisting of both members being early-type galaxies. Our final

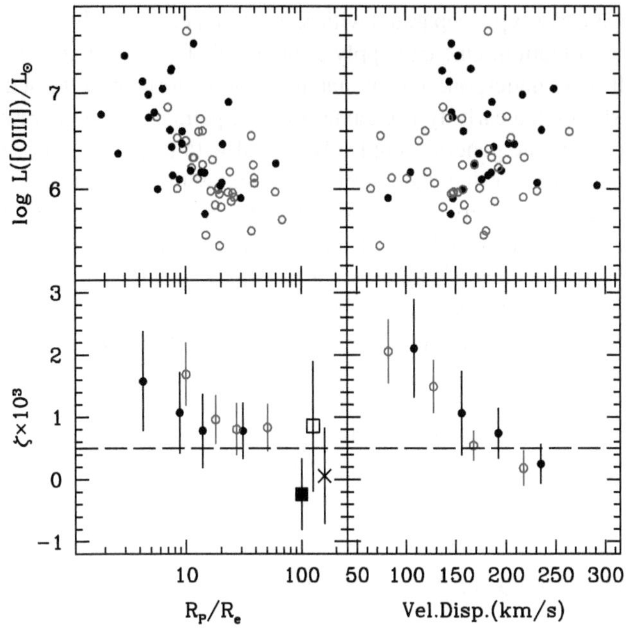

Fig. 1 Spectral properties of the close pairs sample, given by the [OIII] luminosity (*top*) and PCA parameter ζ, that tracks the underlying stellar populations (*bottom*). Values of $\zeta > 500$ (*horizontal dashed line*) reflect a significant amount of star formation. The *black solid* and *grey open dots* represent galaxies with (without) visual signs of interaction (as seen on the SDSS DR6 images). The dots in the *top panel* represent individual galaxies (where AGN-like emission is found), whereas the *bottom panels* give the average and error on the average for subsamples binned according to pair separation (R_P, *left*), given in units of the effective radius) or velocity dispersion (*right*). The *solid* and *hollow squares* in the *bottom-left panel* shows the average and RMS of a field distribution of NUV faint and bright early-type galaxies, respectively, and the *cross* shows the average value, regardless of NUV flux

sample comprises 347 pairs, and is described in [9]. This definition of a pair produces a clean sample to probe the details of the triggering of star formation and the onset of nuclear emission, as the progenitors are expected to have neither star formation nor AGN activity. The top panels of Fig. 1 shows the luminosity of the [OIII] line for galaxies with AGN-like emission (which amounts to 9.5% of the total sample). A trend is found with respect to pair separation (*left*, given in units of the effective radius of each galaxy), but no correlation is found with internal velocity dispersion (*right*). The bottom panels show the distribution of the ζ PCA-component, which tracks recent star formation. In this case, the general sample (not only AGN) is shown, with the dots and error bars representing the average and its error on subsamples binned according to separation or velocity dispersion. The decomposition into PCA components is explained in detail in [8]. As a comparison, the squares in the bottom left panel give the values of ζ for a field sample of early-types separated with respect to their NUV flux – a direct tracer of recent star formation. One can apply a threshold at $\zeta > 500$ above which most galaxies will have undergone recent star formation (horizontal dashed line). The comparison shows that early-type galaxies in close pairs have a higher rate of recent star formation than the general sample. The trend with respect to pair separation is visible, although weaker than the correlation found with respect to a more intrinsic observable, namely velocity dispersion. The observed onset of star formation in close pairs, even before the galaxies display any visual feature of interaction, can be explained by the presence of clouds of gas within their halos, as observed by, e.g. [2]. Even in galaxies separated over 100 effective radii, we find that their level of recent star formation is higher than the average value found for a general sample of (non close pair) early-type galaxies, shown by the '×' sign on the bottom-left panel of Fig. 1. Furthermore, the strong correlation between pair separation and AGN activity – traced by the luminosity of the [OIII] line – suggests that the triggering of star formation precedes AGN activity (see [9] for details).

3 Environment Over Large Scales: Groups

For a more general description of environment, we use a sample of SDSS early-type galaxies from [1] and define their environment according to the host dark matter halo, given in the groups catalogue of [11]. This definition of 'group' is more general than the traditional concept of galaxy groups, and can extend from isolated galaxies – where one dark matter halo contains only one galaxy, to clusters. However, our sample treats galaxies within a group in the same way, so that we do not resolve, for instance, the environmental differences between the central region and the outskirts of a galaxy cluster. Rather, we want to quantify the importance of the background density where the galaxy lives, on its past and recent star formation history (see [10]).

Each galaxy is described by a 'local' observable (their central velocity dispersion) and by an 'environmental' parameter (the host halo mass). Figure 2 shows the trend of the PCA-based parameters (η and ζ, with the subindex '3' meaning $\times 10^3$), with respect to the local and environmental observables. The shaded areas map the RMS of the distribution, segregated with respect to halo mass or velocity dispersion, as labelled. For clarity, the black shaded area in the top panels is replaced by solid and dashed black lines, tracing the mean and RMS, respectively. The bottom panel shows the fraction of galaxies with a value of ζ above the 500 threshold for which the galaxy is assumed to have undergone recent star formation (see the comparison in the bottom left panel of Fig. 1 between ζ and NUV bright/faint galaxies). The figure suggests local properties, such as velocity dispersion, play an important role in shaping the underlying stellar populations, both for the average age (η, top panels) and for the amount of recent star formation (ζ, bottom panels). Environment, as defined here by halo mass, only gives a correction to this trend, with galaxies in less massive halos appearing with a similar average age, but with an slightly higher

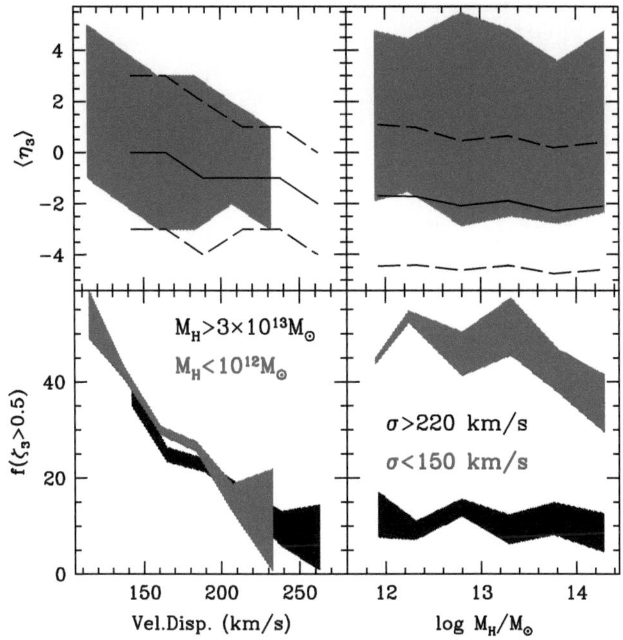

Fig. 2 Effect of group environment on the stellar populations of early-type galaxies. The *top panels* show the value of PCA parameter $\eta_3 \equiv \eta \times 10^3$, which maps average stellar age. The *bottom panels* give the fraction of galaxies with recent star formation, as characterized by the PCA parameter $\zeta_3 \equiv \zeta \times 10^3 > 0.5$. The samples are shown with respect to a local property (velocity dispersion, *left*) and an environment-related property (group halo mass, *right*)

amount of recent star formation at fixed velocity dispersion. We should emphasize here that this definition of environment is adequate for large scale structure analyses. The more acute effects of environment, such as those encountered when galaxies fall into the potential wells of clusters, are not accounted for in this definition.

References

1. M. Bernardi et al., AJ **131**, 1288 (2006)
2. F. Combes, L.M. Young, M. Bureau, MNRAS **377**, 1795 (2007)
3. C.J. Conselice, MNRAS **399**, L16 (2009)
4. A. Dressler, ApJ **236**, 351 (1980)
5. I. Ferreras et al., MNRAS **370**, 828 (2006)
6. I. Ferreras et al., MNRAS **396**, 1573 (2009)
7. A. Hyvärinen, J. Karhunen, E. Oja, *Independent Component Analysis* (Wiley, NY, 2001)
8. B. Rogers et al., MNRAS **382**, 750 (2007)
9. B. Rogers et al., MNRAS **399**, 2172 (2009)
10. B. Rogers et al., MNRAS **405**, 329 (2010)
11. X. Yang et al., ApJ **671**, 153 (2007)
12. D.G. York et al., AJ **120**, 1579 (2000)

Revealing the Origins of S0 Galaxies Using Maximum Likelihood Analysis of PNe 2D Kinematics: The Role of Environment

A. Cortesi, M. Merrifield, M. Arnaboldi, and The PN.S consortium

Abstract Lenticular galaxies lie between ellipticals and spirals on the Hubble sequence, since they have the featureless old stellar populations of elliptical systems, but also contain the disk components associated with spirals. At the moment there is no consensus as to the end of the Hubble sequence to which they are more closely related. Clearly, some process has shut off their star formation; they could be the result of galaxy mergers, much in the manner that star formation is believed to be quenched in elliptical galaxies, or they could be relatively normal spiral galaxies whose star formation has been stopped by some gentler process. The best way of discriminating between these scenarios is offered by the stellar dynamics of S0 galaxy disk. We present a new method based on a maximum likelihood analysis and a bulge-to-disk light decomposition to recover the star kinematic of lenticular galaxies. We test the method on the galaxy NGC 1023.

1 Kinematic Likelihood Fitting: Method and Results

To investigate the origins of S0 galaxies, we present a new method for analyzing their stellar kinematics with discrete tracers such as planetary nebulae. This method involves binning the data in the radial direction so to extract the most general possible non-parametric kinematic profiles, and using a maximum likelihood fit within each bin in order to make full use of the information in the discrete

A. Cortesi (✉)
School of Physics and Astronomy, University Park, Nottingham NG7 2RD, UK
e-mail: acortesi@eso.org

M. Merrifield
Centre for Astronomy and Particle Physics, University Park, Nottingham NG7 2RD, UK

M. Arnaboldi
European Southern Observatory, Karl-Schwarzschild-Str. 2, 85748 Garching, Germany

I. Ferreras and A. Pasquali (eds.) *Environment and the Formation of Galaxies: 30 years later*, Astrophysics and Space Science Proceedings,
DOI 10.1007/978-3-642-20285-8_21, © Springer-Verlag Berlin Heidelberg 2011

kinematic tracers. Both disk and spheroid kinematic components are fitted, with a two-dimensional decomposition of imaging data used to assign each tracer a probability of membership in the separate components. Likelihood clipping also allows us to identify objects whose properties are not consistent with the adopted model, rendering the technique robust against contaminants and able to identify additional kinematic features.

We first tested the method on a N-body simulated galaxy to assess possible sources of systematic error associated with the structural and kinematic decomposition, which were found to be small. It was then applied to the S0 system NGC 1023, for which a PN catalog has already been released and analyzed, see [3]. The correct inclusion of the spheroidal component allows us to show that, contrary to previous claims, the stellar kinematics of this galaxy are indistinguishable from those of a normal spiral galaxy, with rotation velocity higher than random motions at large radii, as shown in Fig. 1. This indicates that NGC 1023 may have evolved directly from a spiral galaxy via gas stripping or secular evolution. The method also successfully identifies a population of outliers whose kinematics are different from those of the main galaxy; these objects can be identified with a stellar stream associated with the companion galaxy NGC1023A.

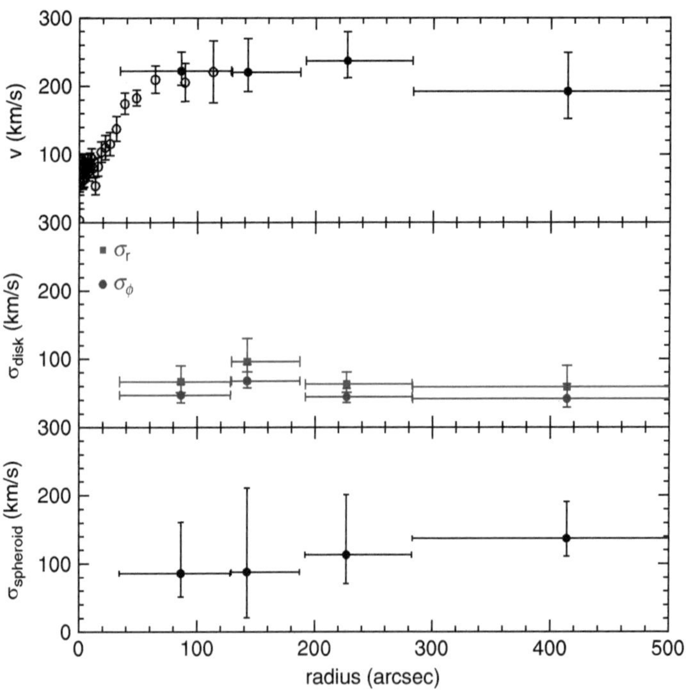

Fig. 1 The *filled symbols* correspond to the parameters recovered with the likelihood methods, with 1σ error bars. The *open symbols* show rotation velocities derived from absorption line data by [1]

The next step in this project will involve applying this analysis technique to observations of a larger sample of S0 galaxies whose PNe kinematics have been observed with The Planetary Nebula Spectrograph [PN.S, [2]], to investigate whether there is a single route to S0 formation irrespective of environment.

References

1. V.P. Debattista et al., MNRAS **332**, 65 (2002)
2. N.G. Douglas et al., PASP **114**, 1234 (2002)
3. E. Noordermeer et al., MNRAS **384**, 943 (2008)

Stellar Populations in the Outskirts of M33

M. Grossi, N. Hwang, E. Corbelli, C. Giovanardi, S. Okamoto, and N. Arimoto

Abstract We present V and I photometry of seven fields in the outskirts of M33 to search for stellar structures in correspondence of neutral hydrogen (HI) clouds recently discovered in a 21-cm survey of this galaxy [2]. We analyse the stellar population out to 2° from the centre of M33 by means of V, I colour magnitude diagrams reaching the red clump (RC). We detect intermediate-age (1–10 Gyr) stars out to a deprojected radius of 30 kpc (~20 scale lengths), but we do not find evidence for a clear correlation between the HI complexes and the stellar substructures.

1 Data Set

Observations were obtained with Suprime-Cam on the 8.2-m Subaru telescope. The regions observed during the run are shown in Fig. 1. The fields are labeled from 1 to 7 and they are related to varied HI features around the galaxy [2]. The total exposure times were 2,200 and 4,800 s in V and I, respectively. The colour magnitude diagrams (CMD) of the target fields are shown in Fig. 2, with isochrones from [1] overlaid to highlight the mean age and metallicity of the stellar populations in each field.

M. Grossi (✉)
CAAUL, Observatório Astronómico de Lisboa, Universidade de Lisboa, Lisbon, Portugal

N. Hwang · S. Okamoto · N. Arimoto
Department of Astronomical Science, Graduate University of Advanced Studies, Tokyo, Japan

E. Corbelli · C. Giovanardi
INAF-Osservatorio Astrofisico di Arcetri, Florence, Italy

I. Ferreras and A. Pasquali (eds.) *Environment and the Formation of Galaxies: 30 years later*, Astrophysics and Space Science Proceedings,
DOI 10.1007/978-3-642-20285-8_22,

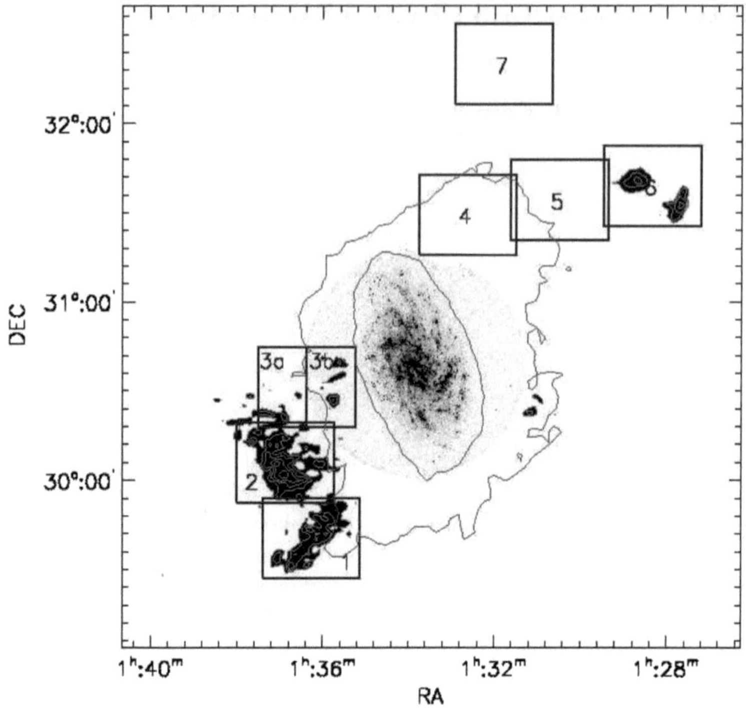

Fig. 1 Location of the seven target fields around M33 imaged with SUBARU/Suprime-Cam. The two main HI features to the south–east and north–west of the galaxy are also shown. The M33 HI disc is highlighted by two contours with a column density of $N_{HI} = 5$ and $50 \times 10^{19}\,cm^{-2}$, respectively

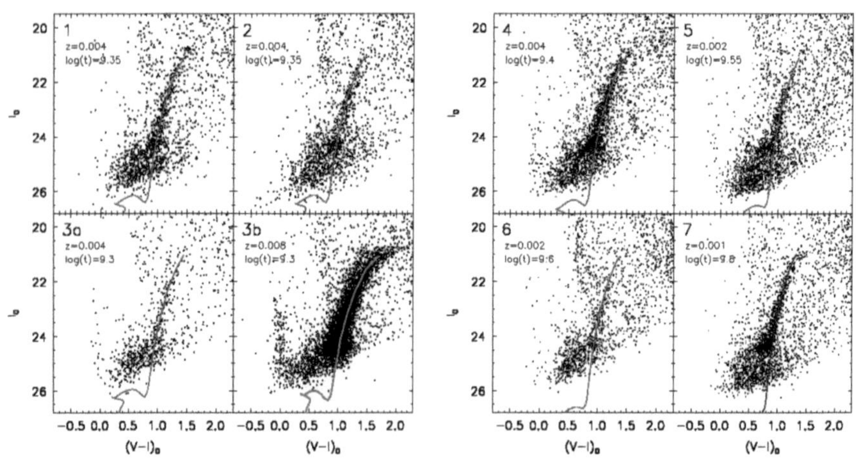

Fig. 2 *Left:* CMDs of target fields. The most prominent feature is the red giant branch (RGB), corresponding to stars with colours $(V - I)_0$ between 0.7 and 2 mag, and $I_0 < 24$, which is detected out to the most distant fields from the M33 centre. Isochrones from [1] are overlaid to give an indication of the mean age and metallicity of the RGB/RC stars

2 Results

1. We detected stellar structures out to the most distant fields from the M33 center, although there is no clear evidence for an optical counterpart to the main HI complexes. The northernmost field (n. 7) shows a peculiar enhancement in the number density of RGB and RC stars, given that the corresponding deprojected radius is ~30 kpc.
2. A diffuse population of young stars is detected both to the north and to the south out to a projected distance of ~20 kpc (10 scale lengths).
3. We used the RGB and RC features of the CMDs to derive the mean age and metallicity of the stellar populations in each field. The mean metallicity in the southern fields remains approximately constant outside the edge of the optical disc, at $Z = 0.004$. To the north, we find a clear metallicity gradient (from $Z = 0.004$ to $Z = 0.001$) and age gradient from the inner to the outer fields.

References

1. L. Girardi et al., A&A **391**, 195 (2002)
2. M. Grossi et al., A&A **487**, 161 (2008)

Are Boxy/Disky Ellipticals Dependent on Environment?

B. Häußler, M. Gray and STAGES team

Abstract Boxy and disky isophotes in elliptical galaxies are an imprint of their formation history. It is thus interesting to examine the boxyness of early-type galaxies as a function of environment in large cluster systems, especially as some N-body simulations also predict a higher number of disky versus boxy ellipticals in overdense regions. Using high-resolution *HST* data from the STAGES survey, centred on the Abell901/902 supercluster at a redshift of $z \sim 0.167$, we investigate this dependence. Taking extreme care of the setup of the codes used (particularly `ellipse`), we analyse a sample of ~560 ellipticals and S0 galaxies, one of the biggest samples examined for this effect. Although other groups have found environmental dependencies before, no such dependence is found in the STAGES survey data. The ratio of boxy to disky galaxies stays constant over the entire range of environment and galaxy density present in the field examined.

1 Introduction

Since the isophotal shapes of nearby galaxies were studied to examine the formation history of galaxies (e.g. [1–3]) e.g. in the Virgo cluster, people found differences between the populations of disky and boxy ellipticals, e.g. correlations of isophotal shape with radio-brightness (radio-loud galaxies are more likely to be boxy) or gas content (galaxies with gaseous X-ray halos are more boxy). Also, boxy ellipticals show higher M/L ratios than ellipticals with weak disks. All this hints towards different formation scenarios between the different populations which can be backed up using N-body simulations, e.g. [8]. The authors of [7] shows that equal mass

B. Häußler (✉) · M. Gray
School of Physics and Astronomy, University of Nottingham, University Park, Nottingham NG7 2RD, UK
e-mail: Boris.Haeussler@nottingham.ac.uk; Meghan.Gray@nottingham.ac.uk

I. Ferreras and A. Pasquali (eds.) *Environment and the Formation of Galaxies: 30 years later*, Astrophysics and Space Science Proceedings,
DOI 10.1007/978-3-642-20285-8_23,

mergers of spiral galaxies and major mergers of ellipticals form boxy isophotes, whereas un-equal mass mergers that include spiral galaxies (spiral-elliptical or spiral-spiral) form disky isophotes in elliptical galaxies.

Most of the previous work on isophotal shapes has been done both on small galaxy samples, where taking care of individual peculiarities of the objects is feasible, and on local galaxies, where the data resolution is good, the details of the galaxy structure are visible and S/N is high. Here, we present an automated approach to the fitting of galaxy shapes on a large sample of ∼560 galaxies at $z \sim 0.167$.

2 Data, Sample Selection and Data Processing

In this work, we use galaxy catalogues provided by the STAGES survey [4] that contain both photometric information from COMBO-17 [11] and profile information from galaxy profile fitting using GALAPAGOS (Barden et al. in preparation; also see [5]), which uses GALFIT [10] and a complicated method for deblending/masking, background estimation and simultaneous fitting to derive the best fitting result possible. STAGES further provides high-resolution *HST/ACS* images that cover the Abell 901/902 supercluster at $z \sim 0.167$ that are needed for fitting the isophotal shapes of galaxies. With sky coverage of 1/4 deg^2 it is one of the biggest *HST* surveys in existence, and forms one of the largest samples of galaxies for this kind of work.

From this publicly available catalogue of 88,879 galaxies we select galaxies that have been independently visually classified by several members of our team (Gray 2011, in preparation). In total, 5,009 galaxies down to $Rmag \sim 23.5$ have been selected for classification out of which we select those classified as E or S0 to derive a pure morphologically selected sample instead of using ambiguous values like colour or galaxy profile. We use a magnitude dependent redshift cut, to divide this sample into cluster and fields samples.

To be able to derive good fitting values for each galaxy when using the `IRAF` task `ellipse` [6], we clean all images of potential neighbours using GALAPAGOS derived galaxy profiles and check all images for artificially introduced artefacts (e.g. sky edges,...). Some of the cleaned image postage stamps of galaxies in our sample are shown in Fig. 1. We then use `ellipse` to derive isophotal parameters for all galaxies. We choose a setup with logarithmic stepping and which allows the centring of the individual isophotes to vary with radius (for more details see Häußler et al., 2011, in preparation). Running the code this way, it still happens for a fraction of the objects, that the fit returns a stop code `STOP`$\neq 0$ for at least some of the isophotes within a galaxy (see Fig. 2) as the starting values for ellipticity and position angle were chosen arbitrarily. This is especially critical for isophotes around our starting value of the fit at radius 10 pixels as this shows that the fitting results were sensitive to our starting guesses. In the cases where this happens, we rerun the code, now starting at an interpolated value for both ellipticity and position angle. The continuous profiles in the last row of Fig. 2 and the improvement in the

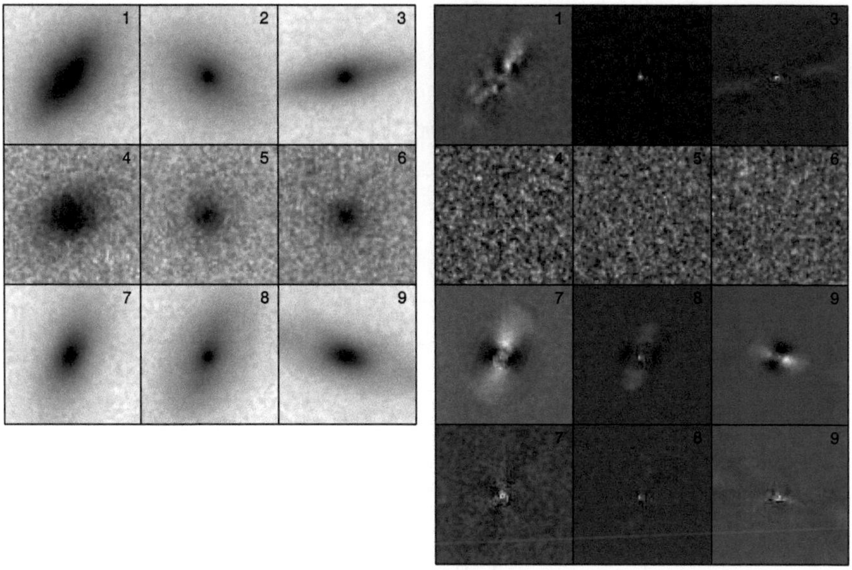

Fig. 1 The *left* images show some of the galaxies in our sample, all are clearly early-type galaxies at different noise levels. The plot on the *right* shows the fitting residuals where the ellipse model has been subtracted. Especially for the second row of objects, it is interesting to see that, although ellipse returns non-ideal fits for a fraction of the annuli during the fit (see Fig. 2), the fitting residuals look remarkably empty. In case of the objects in the last row, two different residuals are shown for the two different runs of ellipse

fitting residuals in Fig. 1 show the success of this approach. In total, this second step fitting was necessary for around 25% galaxies. After visual inspection of all the fitting residuals, the vast majority of the automatic decisions were approved, only five had to be overruled as the first fit clearly showed the smaller residual. By visual inspection of the galaxy images, we cannot find anything particularly strange about these objects, but they are all quite small, faint objects for which any fit is possibly very challenging in the first place.

From the complete radial profile derived in this way, we get a global value that is characteristic for the whole galaxy as individually comparing hundreds of full profiles is simply not possible in most works. Similar to [9] and [2], we choose an intensity weighted mean value for this and the width of the distribution as its uncertainty. To avoid effects of the PSF (at small radii) or large uncertainties due to noise in the rings (at very large radii), we select valid radii using STOP $= 0$ and $5[pixels] \leq r \leq 1.5r_e$ with 5 pixels being roughly 2 FWHM of the STAGES PSF and r_e being the half-light radius as derived by GALAPAGOS (see Fig. 3 for details).

Our final catalogue contains a total number of 560 galaxies that split up into 100 field ellipticals, 188 cluster ellipticals, 59 field S0 and 213 cluster S0 galaxies. Throughout our analysis, we count B4 as being different to 0 if the mean is offset by more than one σ, as otherwise even clear visual offsets, e.g. in Fig. 3 in the upper

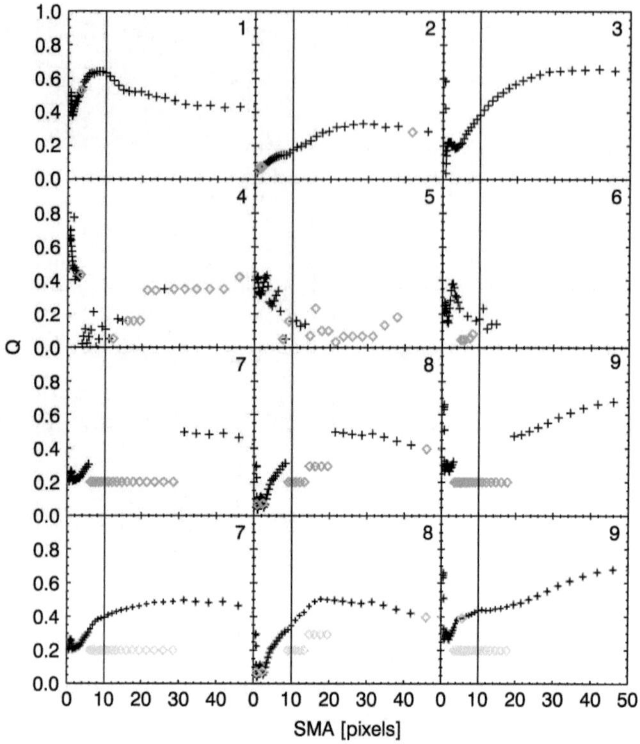

Fig. 2 Ellipticity profiles as derived by ellipse fitting for the galaxies shown in Fig. 1. *Black crosses* show valid values as returned by ellipse, *grey squares* show these with *stop* > 0. For the galaxies in the *top row* ellipse immediately returns good results, the second shows profiles of objects for which the code has difficulty at certain radii, but not at our starting value of 10 pixels. For the objects in the third row, our first attempt fails at the starting radius of 10 pix (see the plateaus around that value). In the last row, we show the results for these galaxies when starting at interpolated values. We plot ellipticity here rather than B4 to show the effects more clearly

left panel, would be discarded. With $B4 > 0$, a galaxy shows disky isophotes, with $B4 < 0$, it shows boxy isophotes.

3 Results and Comparison

Although we can see clear differences between the sample of ellipticals and the comparison sample of S0 galaxies, which shows that our boxyness/diskyness (B4) measurements are good measures of the real values, we see no significant effect when we look at the dependence of the characteristic B4 values of ellipticals as a function of environment, independent of the measure for environment used (see Fig. 4). This leads us to the conclusion that if this effect is expected and real, it

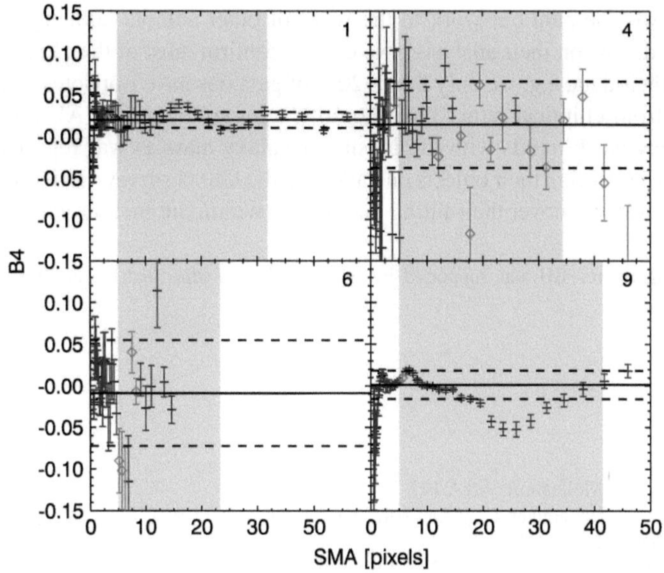

Fig. 3 At the example of boxyness/diskyness B4, we show how the characteristic values are derived. *Black points* show radii with valid values as returned by ellipse, grey boxes show ignored radii. The *light grey shaded area* shows the area of radius, $5pix < r < 1.5r_e$, that is used to derive the characteristic value shown as the horizontal *solid line*, the *dashed lines* show $\pm 1\sigma$. To connect the galaxies to the other plots, they are labelled with 1, 4, 6 and 9 as given in Fig. 1

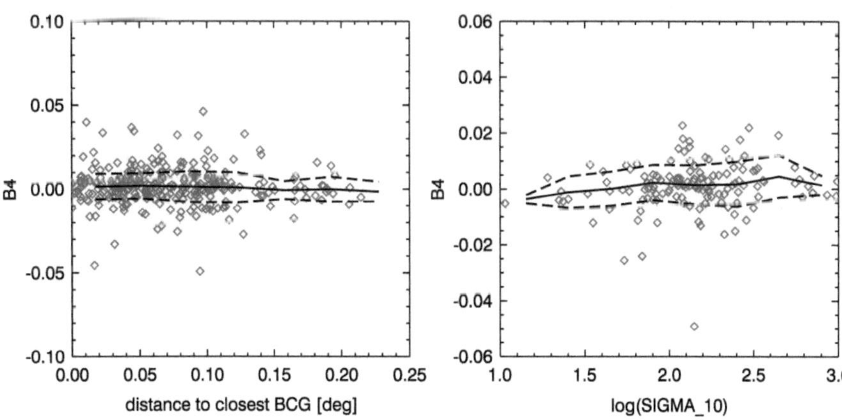

Fig. 4 This plot shows the boxyness/diskyness of elliptical galaxies as a function of apparent distance to the nearest BCG galaxy (*left panel*) and σ_{10} (*right panel*) as derived from STAGES data. In neither of the plots, a significant trend can be seen

is possibly a very subtle effect which is well hidden in our measurement errors or measure of environment or that it is not observable in the range of environment present in the STAGES data, but relies on more extreme environments, e.g. centres of Coma-like clusters. From our work, no immediate conclusion can be drawn concerning the assembly history of the elliptical galaxies in the STAGES field.

When we compare our work to the work of other authors (although no author has given details on their analysis), we cannot confirm most of the effects that they found in their data, e.g. we only find ~20% of galaxies have isophotes significantly different from elliptical rather than the 60% reported by [2, 3]. Also we can not confirm any weak trends with luminosity or galaxy mass as was reported by [9], although when fitting their objects on data from the GEMS survey, similar to what we use here, we can recover their fitting parameters within our measurement errorbars.

Acknowledgements BH was supported by STFC, MG was supported by an STFC Advanced Fellowship.

References

1. R. Bender, C. Möllenhoff, A&A **177**, 71–83 (1987)
2. R. Bender, S. Döbereiner, C. Möllenhoff, ApJ **74**, 385–426 (1988)
3. R. Bender et al., A&A **217**, 35–43 (1989)
4. M.E. Gray et al., MNRAS **393**, 1275–1301 (2009)
5. B. Häussler et al., ApJS **172**, 615–633 (2007)
6. R.I. Jedrzejewski, MNRAS **226**, 747–768 (1987)
7. S. Khochfar, A. Burkert, MNRAS **359**, 1379–1385 (2005)
8. T. Naab, S. Khochfar, A. Burkert, ApJL **636**, L81–L84 (2006)
9. A. Pasquali et al., ApJ **636**, 115–133 (2006)
10. C.Y. Peng et al., Astronomical J, **124**, 266–293 (2002)
11. C. Wolf et al., A&A **421**, 913–936 (2004)

An Environmental Butcher–Oemler Effect in Intermediate Redshift X-Ray Clusters

S. Urquhart and J. Willis

Abstract We present uniform CFHT Megacam g and r photometry for 34 X-ray selected galaxy clusters drawn from the X-ray Multi-Mirror (XMM) Large Scale Structure (LSS) survey and the Canadian Cluster Comparison Project (CCCP) having X-ray temperatures of $1 < \mathrm{kT(keV)} < 12$, and a relatively narrow redshift interval of $0.15 < z < 0.41$ in order to minimize any redshift dependent photometric effects. We investigate the colour bimodality of the cluster galaxy populations and compute blue fractions using criteria derived from [Butcher and Oemler, ApJ 285, 426 (1984)]. We identify an environmental dependence of cluster blue fraction in that cool (low mass) clusters display higher blue fractions than hotter (higher mass) clusters. The results of blue fraction calculations as a function of this global cluster environment alone are not enough to disentangle the process(es) which are transforming galaxies, and so are combined with an initial analysis of local density measurements.

1 Introduction

The observed properties of galaxy populations reflect the environment in which they are located. Comparisons of galaxy populations drawn from low-density (the field) and high-density (rich galaxy clusters) environments indicate that the population distribution described using measures such as current star formation rate (e.g. [1, 14]), integrated colour (e.g. [3]) and morphology (e.g. [7, 15]) varies as a function of changing environment. From studies such as these it is clear that galaxies

S. Urquhart (✉) · J. Willis
University of Victoria, Victoria, BC, Canada
e-mail: surquhar@uvic.ca; jwillis@uvic.ca

I. Ferreras and A. Pasquali (eds.) *Environment and the Formation of Galaxies: 30 years later*, Astrophysics and Space Science Proceedings,
DOI 10.1007/978-3-642-20285-8_24, © Springer-Verlag Berlin Heidelberg 2011

located in the cores of rich clusters display lower star formation rates, redder colours and more bulge-dominated morphologies compared to galaxies located in the field. Studies of the fraction of blue galaxies contributing to a galaxy cluster provided some of the first direct evidence for the physical transformation of galaxies in cluster environments. [4] reported an increase in the fraction of blue galaxies out to z $\sim$ 0.5 compared to local clusters. An alternative approach, employed here, is to consider an environmental Butcher–Oemler effect whereby one attempts to determine the variation of blue fraction as a function of varying intrinsic cluster properties selected over narrow redshift intervals. This has been done previously by comparing blue fractions within clusters at increasing clustercentric radii [8] or by considering blue fractions within clusters of differing X-ray luminosities (e.g. [17]). There are numerous explanations for the physical processes which are transforming galaxy populations. The two principle examples are ram pressure stripping [9] and galaxy-galaxy interactions. When referring to such interactions, this may include tidal encounters (e.g. [5]), harassment [11] or the merging with existing cluster members (e.g. [6]). However, it remains unclear as to whether more than one physical process acts upon a galaxy falling into a dense environment (and which would be considered dominant). Therefore, what is presented here is an analysis of both global and local cluster environments to try and disentangle these effects.

2 The Galaxy Cluster Sample

The data presented are drawn from two complimentary samples of X-ray-selected galaxy clusters. Clusters with X-ray temperatures T$<$ 3 keV and $0.25 < z < 0.35$ are drawn from the X-ray Multi-Mirror (XMM) Large Scale Structure (LSS) survey [12] generating a sample of 11 clusters which are referred to as "cool" in the analysis.

Clusters with X-ray temperatures T$>$5keV are selected from the sample of Horner [10] and form part of the Canadian Cluster Comparison Project (CCCP; [2]). We consider here 27 systems with accompanying optical data obtained using the Canada-France-Hawaii Telescope (CFHT) Megacam imager. Due to the large temperature range covered by the CCCP sample (5 $<$ T(keV) $<$ 12) we further sub-divide the CCCP sample into "mid" clusters displaying 5$<$T(keV) $<$ 8 and "hot" displaying T $>$8 keV. We limit the CCCP sample to redshifts $0.15 < z < 0.4$ in order to maximise the available sample size while reducing the redshift interval over which photometric quantities are k-corrected to a common epoch. These considerations limit the size of the mid and hot samples to 18 and 5 clusters respectively. The characteristic cluster scaling radii employed in this work are based upon r_{500}. For full details of the cluster sample and subsequent blue fraction analysis, please refer to [16].

3 Blue Fractions

Computation of the blue fraction first requires the red sequence relation to be determined as a reference value relative to which the blue galaxy population may be defined, here, considering only sources within r_{500}. Following the statistical background subtraction method of [13], the red sequence location was computed employing a weighted, linear least-squares fit.

The blue fraction of each cluster was calculated following the definition of [4] whereby all galaxies having $M_V \leq -20$ are considered and blue galaxies are defined as those displaying a rest-frame colour offset of $\Delta(B-V) = -0.2$ measured relative to the red sequence. The blue fraction is then given by

$$f_B = \frac{N_{Blue,Total} - AN_{Blue,Field}}{N_{Total} - AN_{Field}}, \tag{1}$$

where $N_{Blue,Total}$ is the number of blue galaxies in the cluster plus field, $N_{Blue,Field}$ is the number of blue field galaxies, N_{Total} is the total number of galaxies in the cluster plus field and N_{Field} is the total number of field galaxies. The symbol A denotes an areal scaling factor to correct the field population area to that of the cluster area.

Figure 1 displays the blue fraction as a function of cluster X-ray temperature for all clusters. The data indicates that typical blue fractions in each cluster and the dispersion in blue fraction values within a given temperature subsample increase as the temperature of the X-ray cluster decreases.

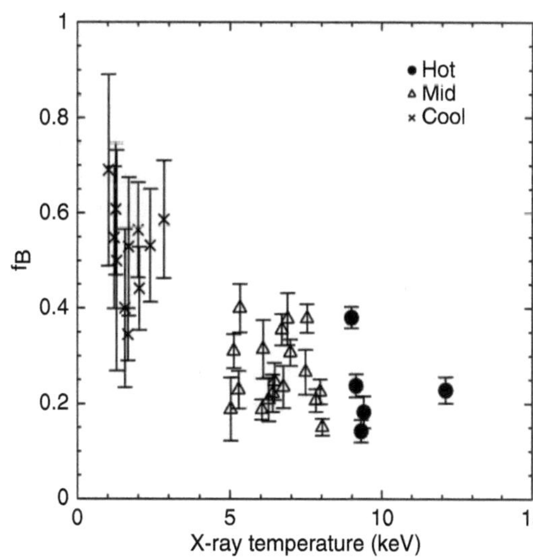

Fig. 1 Cluster blue fraction as a function of X-ray temperature

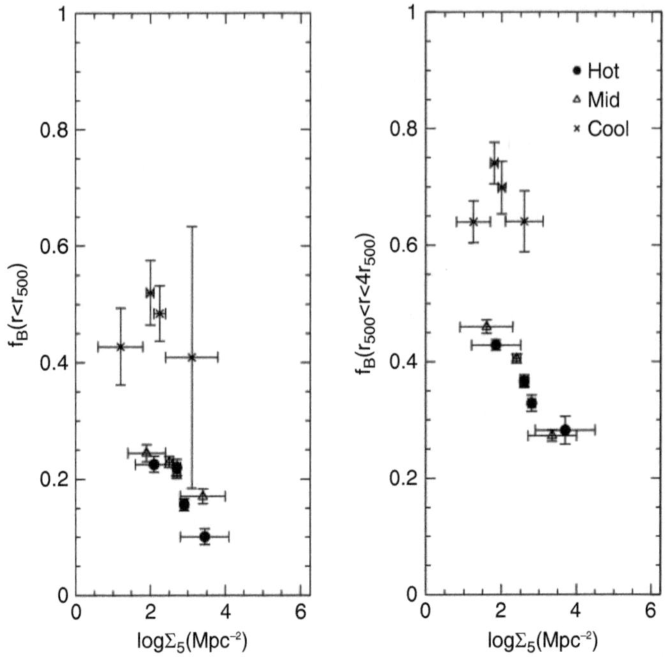

Fig. 2 Blue fraction as a function of local density. The *left panel* shows galaxies within $r < r_{500}$, the *right panel* shows galaxies having $r_{500} < r < 4r_{500}$

4 Local Density

Having considered the effects of global environment on galaxy populations (i.e. investigating the variation of cluster blue fraction as a function of temperature), a local density parameter is now being considered. This is given by Σ_5, an estimate of the density based on the distance to the fifth nearest galaxy from the galaxy of interest, defined as

$$\Sigma_5 = \frac{5}{(\pi r_5^2)}, \qquad (2)$$

where r_5 is the distance to the fifth nearest neighbour from the galaxy of interest.

Based upon the interquartile distances of the appropriate $\log \Sigma_5$ distributions, the data was binned and the blue fractions for each were calculated as shown in Fig. 2. In addition to this, a division was made between "inner" (galaxies having $r < r_{500}$) and "outer" (galaxies having $r_{500} < r < 4r_{500}$) since there is an observed sharp decline in $\log \Sigma_5$ in the inner $r < r_{500}$ region and then a decreasing of the gradient to zero for all clusters. A clear trend can be seen between regions of differing radial cuts. As one travels towards the cluster centre, for clusters of all temperatures, the blue fractions decrease. In addition to this, within a radial cut, in the mid and hot samples, there is a clear increase in blue fraction with a decrease in local density. Both of these trends are nominally consistent with the effects of ram pressure stripping.

5 Discussion

The trend of decreasing blue fraction with increasing X-ray temperature identified here is consistent with an environmental component to the Butcher–Oemler effect. A currently favoured explanation for this is that infalling field galaxies are processed physically as they interact with the cluster environment. This may be due to ram pressure stripping (whose effectiveness is a simple function of mass scale of the cluster) or due to galaxy–galaxy interactions. Distinguishing between these effects is difficult since, to first order, the effects upon star formation in the infalling galaxy depend on environment in the same manner: Intra-cluster medium (ICM) stripping and subsequent star formation suppression are expected to be weakest in groups, whereas star formation enhancement due to interactions is expected to be greater on these group scales. Each of these effects would naively generate the same observed trend in blue fraction versus temperature.

Neither on the basis of the colour distribution in each sample, nor by dividing the analysis into global and local components allow the cause of this environmental Butcher–Oemler effect to be discerned unambiguously. In addition to this, the effects of redshift evolution also require investigation to determine their importance. Clearly, further information is required. This includes extending the redshift interval studied to determine the extent to which such redshift evolutionary effects may influence galaxy populations as well as combining this colour work with morphological studies. Both of these investigations will be presented in two future papers.

Acknowledgements The authors wish to thank Graham Smith, Michael Balogh, Stefano Andreon, Pierre-Alain Duc, Florian Pacaud and Kevin Pimbblet for providing useful discussions. SAU and JPW acknowledge financial support from the Canadian National Science and Engineering Research Council (NSERC).

References

1. M.L. Balogh et al., MNRAS **527**, 54 (1999)
2. C. Bildfell et al., MNRAS **389**, 1637 (2008)
3. M.R. Blanton et al., ApJ **629**, 143 (2005)
4. H. Butcher, A. Oemler, ApJ **285**, 426 (1984)
5. A. Chung et al., ApJ **659**, 115 (2007)
6. T.J. Cox et al., MNRAS **384**, 386 (2008)
7. A. Dressler et al., ApJ **490**, 577 (1997)
8. E. Ellingson et al., ApJ **547**, 609 (2001)
9. J.E. Gunn, J.R. Gott ApJ **176**, 1 (1972)
10. D. Horner, PhD thesis, University of Maryland (2001)

11. B. Moore et al., Nature **379**, 613 (1996)
12. F. Pacaud et al., MNRAS **382**, 1289 (2007)
13. K. Pimbblet et al., MNRAS **331**, 333 (2002)
14. B.M. Poggianti et al., ApJ **642**, 188 (2006)
15. T. Treu et al., ApJ **591**, 53 (2003)
16. S.A. Urquhart et al., MNRAS **406**, 368 (2010)
17. D.A. Wake et al., ApJ **627**, 186 (2005)

Testing the Hierarchical Scenario with Field Disk Galaxy Evolution

A. Böhm and B.L. Ziegler

Abstract We have constructed a data set of >250 field disk galaxies at redshifts $0.1 < z < 1.0$ with Very Large Telescope (VLT) spectroscopy and Hubble Space Telescope imaging. This is one of the largest kinematical samples of distant disks to date. We use spatially resolved rotation curves to derive maximum rotation velocities and total masses; we also investigate disk sizes, bulge-to-disk ratios, stellar population properties etc. The stellar-to-total-mass ratios are constant over the probed cosmic epoch, which favors a hierarchical buildup of the dark matter halos the disks reside in. On the other hand, the mean stellar mass-to-light ratios evolve more strongly in the low-mass galaxies than in high-mass galaxies and the mean stellar ages are lower for low-mass galaxies than for high-mass galaxies. This points to an anti-hierarchical evolution of the stellar populations (aka "downsizing").

1 Our Sample

Our observations were carried out with the FORS instruments of the VLT (spectroscopy) and the Advanced Camera for Surveys of the HST (F814W imaging). Additionally we can rely on deep multi-band optical/NIR photometry. Our measurements of the disk maximum rotation velocity take into account all geometrical and observational effects including seeing and optical "beam smearing" (for details, see [1]). Very recently, we have conducted a survey of very low- and very high-mass disks at redshift $z \approx 0.5$ (PI A. Böhm); these data are among the deepest spectra of distant spiral galaxies ever taken with the VLT. In particular, we will investigate the evolution of the Tully-Fisher Relation and the correlation of maximum rotation velocity with central stellar velocity dispersion (this is work in preparation).

A. Böhm (✉)
Institute of Astro- and Particle Physics, Technikerstrasse 25/8, 6020 Innsbruck, Austria
e-mail: Asmus.Boehm@uibk.ac.at

B.L. Ziegler
Department of Astronomy, University of Vienna, Türkenschanzstraße 17, 1180 Vienna, Austria
e-mail: bodo.ziegler@univie.ac.at

I. Ferreras and A. Pasquali (eds.) *Environment and the Formation of Galaxies: 30 years later*, Astrophysics and Space Science Proceedings,
DOI 10.1007/978-3-642-20285-8_25,

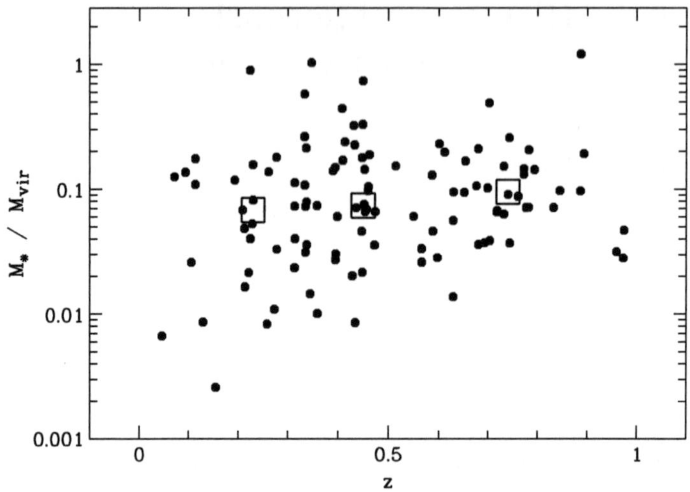

Fig. 1 The fraction between stellar and total mass as a function of redshift. See text for details

2 Results and Discussion

The fraction between the stellar mass M_* and total mass $M_{\rm vir}$ remains roughly constant between redshifts $z = 1$ and $z \approx 0$ (see Fig. 1), which could be understood in terms of smooth accretion of gas and/or Dark Matter-dominated satellites. If spiral galaxies already contained all their dark and baryonic matter at $z \approx 1$, the conversion of gas into stars via star formation would lead to an increase of the stellar mass fraction $M_*/M_{\rm vir}$ towards lower redshifts.

The stellar mass-to-light ratios evolve more strongly for low-luminosity spirals than for high-luminosity spirals [2]. We thus find evidence for an anti-hierarchical evolution of the stellar populations ("down-sizing"). This interpretation gains support from an analysis of the optical and NIR colors with single-zone models: the mean model stellar ages of the distant low-mass spirals are younger than those of the distant high-mass spirals (see [3]). "Down-sizing" so far has mainly been investigated in early-type galaxies, where it possibly is caused by AGN feedback. In quiescent disk galaxies (as in our sample), it might be driven by supernova feedback.

References

1. A. Böhm et al., A&A **420**, 97 (2004)
2. A. Böhm, B.L. Ziegler, ApJ **668**, 846 (2007)
3. I. Ferreras et al., MNRAS **355**, 64 (2004)

The Effect of the Environment on the Gas Kinematics and Morphologies of Distant Galaxies

Y.L. Jaffé

Abstract With the aim of understanding which physical processes are primarily responsible for the transformation of spiral galaxies into S0s in clusters, we study the gas kinematics, morphological disturbances, and the Tully-Fisher relation (TFR) of distant galaxies in various environments. We use the ESO Distant Cluster Survey (EDisCS) dataset, that spans a broad range of cluster and galaxy properties at $0.3 < z < 0.8$. Our results indicate that the physical mechanism acting on cluster galaxies (with $M_B \lesssim -20$ mag) must be strong enough to significantly disturb the gas in cluster galaxies, but at the same time, mild enough to leave the stellar structure unaffected.

Over the past decades, much observational evidence has suggested the transformation of star-forming spirals into passive S0s by the influence of the cluster environment. However, the physical processes responsible for this transformation remain unclear. A number of plausible mechanisms have been proposed. These include, ram-pressure stripping [6], mergers [2], galaxy harassment [11] and tidal stripping [1,8]. Each one of these mechanisms is expected to be effective in different regions of the cluster environment and affect galaxy properties in different ways. For example, a potential key difference between the ram pressure stripping and the merger or tidal stripping scenario is that the former is likely to enhance star formation across the disk [3], while mergers or tides could cause a starburst that is probably centrally concentrated [9]. To attack this problem, we study the effect of environment on the gas kinematics of distant galaxies as well as the morphological disturbances. We focus on a luminosity-limited ($M_B \lesssim -20$ mag) sub-sample of

Y.L. Jaffé (✉)
School of Physics and Astronomy, The University of Nottingham, University Park, Nottingham NG7 2RD, UK
e-mail: ppxyj@nottingham.ac.uk

I. Ferreras and A. Pasquali (eds.) *Environment and the Formation of Galaxies: 30 years later*, Astrophysics and Space Science Proceedings,
DOI 10.1007/978-3-642-20285-8_26,

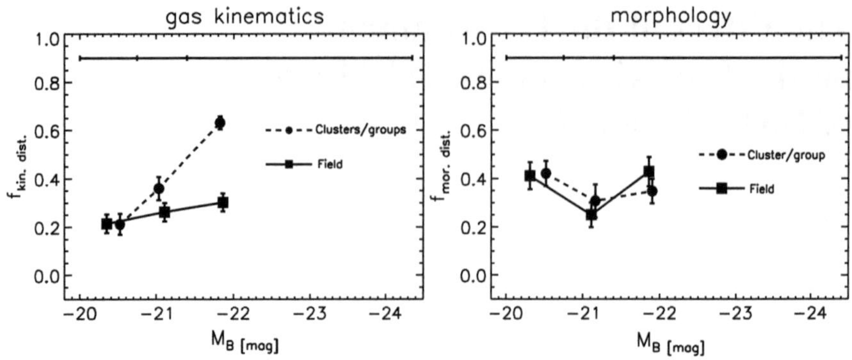

Fig. 1 Fraction of kinematically disturbed (*left*) and morphologically disturbed (*right*) galaxies as a function of M_B for different environments (*labeled*). M_B bins are shown at the *top* of each panel

emission-line galaxies from the EDisCS dataset (see [4,7,10,12]) at $0.3 < z < 0.8$. Our sample consists of 224 spectroscopically confirmed cluster and field galaxies. From the emission-lines in the spectra, we measure rotation velocities and identify galaxies with significant kinematical disturbance. For most of the sample we also have HST imaging from which we are able to quantify morphological disturbances.

Our main result is shown in Fig. 1, where the fraction of kinematically disturbed galaxies (left), and the fraction of galaxies with disturbed morphologies (right) are plotted as a function of rest frame B-band magnitude and environment. It is clear that there are significantly more kinematically disturbed galaxies in cluster environment than in the field, and that the morphological disturbance does not seem to be affected by environment. We interpret the rise of the kinematical disturbance with brighter M_B as a result of smaller cluster galaxies being already stripped off their gas and thus not present in our (emission-line) sample.

We also compare the TFR of cluster and field galaxies (in matched samples of z and M_B) of those galaxies with no signs of kinematical disturbance (i.e. reliable rotation velocities). At face value, we conclude that there are no environmental effects on the TFR, although we are currently investigating the star formation properties of the kinematically disturbed galaxies to test if they could enhance M_B in the TFR.

Finally, we found that although the vast majority of our emission-line galaxies are spirals and a few irregulars, there are also early-type galaxies, some of which have an extended gas disk in their emission. We are currently investigating the origin of the gas in these objects and their link to lower redshift analogs (e.g. [5]).

The results discussed above, and other related results, are discussed in more detail in Jaffé et al. 2010a, b (in preparation)

Acknowledgements Based on observations collected at the European Southern Observatory, Chile, as part of programme 073.A-0216. The author is greatly indebted to the European Southern Observatory, the Royal Astronomical Society and the astrophysics group of Exeter University.

References

1. M.L. Balogh, J.F. Navarro, S.L. Morris, ApJ **540**, 113 (2000)
2. K. Bekki, ApJL **502**, L133+ (1998)
3. K. Bekki, W.J. Couch, ApJ **596**, L13 (2003)
4. V. Desai et al., ApJ **660**, 1151 (2007)
5. E. Emsellem et al., MNRAS **379**, 401 (2007)
6. J.E. Gunn, J.R. Gott, ApJ **176**, 1 (1972)
7. C. Halliday et al., A&A **427**, 397 (2004)
8. R.B. Larson, B.M. Tinsley, C.N. Caldwell, ApJ **237**, 692 (1980)
9. J.C. Mihos, L. Hernquist, ApJ **425**, L13 (1994)
10. B. Milvang-Jensen et al., A&A **482**, 419 (2008)
11. B. Moore, G. Lake, T. Quinn, J. Stadel, MNRAS **304**, 465 (1999)
12. S.D.M. White et al., A&A **444**, 365 (2005)

Deprojecting the Quenching of Star Formation in and Near Clusters

G.A. Mamon, S. Mahajan, and S. Raychaudhury

Abstract Using H_δ and D_n4000 as tracers of recent or ongoing efficient star formation, we analyze the fraction of SDSS galaxies with recent or ongoing efficient star formation (GORES) in the vicinity of 268 clusters. We confirm the well-known segregation of star formation, and using Abel deprojection, we find that the fraction of GORES increases linearly with physical radius and then saturates. Moreover, we find that the fraction of GORES is modulated by the absolute line-of-sight velocity (ALOSV): at all projected radii, higher fractions of GORES are found in higher ALOSV galaxies. We model this velocity modulation of GORES fraction using the particles in a hydrodynamical cosmological simulation, which we classify into virialized, infalling and backsplash according to their position in radial phase space at $z = 0$. Our simplest model, where the GORES fraction is only a function of class does not produce an adequate fit to our observed GORES fraction in projected phase space. On the other hand, assuming that in each class the fraction of GORES rises linearly and then saturates, we are able to find well-fitting 3D models of the fractions of GORES. In our best-fitting models, in comparison with 18% in the virial cone and 13% in the virial sphere, GORES respectively account for 34% and 19% of the infalling and backsplash galaxies, and as much as 11% of the virialized galaxies, possibly as a result of tidally induced star formation from galaxy-galaxy interactions. At the virial radius, the fraction of GORES of the backsplash population is much closer to that of the virialized population than to that of the infalling galaxies. This suggests that the quenching of efficient star formation is nearly complete in a single passage through the cluster.

G.A. Mamon (✉)
IAP (UMR 7095: CNRS and UPMC), Paris, France
e-mail: gam@iap.fr

S. Mahajan · S. Raychaudhury
School of Physics and Astronomy, University of Birmingham, Birmingham, UK

I. Ferreras and A. Pasquali (eds.) *Environment and the Formation of Galaxies: 30 years later*, Astrophysics and Space Science Proceedings,
DOI 10.1007/978-3-642-20285-8_27, © Springer-Verlag Berlin Heidelberg 2011

1 Introduction

It is well known that the cluster environment affects the physical properties of galaxies, as there is a segregation with projected radius of morphology [7], color (e.g. [2]), luminosity (e.g. [1]) and spectral indices such as the equivalent width of [OII] or H_α (e.g. [3, 8]). At the same time, there are indications of velocity segregation of luminosity [4] and of: emission-line galaxies tend to span a wider dispersion of velocities than galaxies without such lines [5]. In fact, it was long known that spiral galaxies in clusters span a wider distribution of velocities than ellipticals and S0s (e.g. [13]). However, those spirals without emission lines and that are not morphologically disturbed span the same velocity distribution as the early type galaxies [12]. Relative to passive galaxies (without H_α in emission), the trend for higher velocity dispersions of galaxies with emission (H_α) lines found by [5] is reversed outside the virial radius [14].

In the present work [10], we take advantage of the large statistics of the Sloan Digital Sky Survey (SDSS) to better quantify the velocity modulation of the radial segregation of the diagnostics of star formation efficiency. We then model the SDSS observed fractions of Galaxies with Ongoing or Recent Efficient Star Formation (GORES), with the help of a cosmological simulation.

2 Observed Velocity Modulation

We select clusters of at least 15 galaxies from the SDSS-DR4 group catalog of [16], with criteria $0.02 \le z \le 0.12$, $M_{180} > 10^{14}\,M_\odot$ and with at least 12 $M_r < -20.5$ galaxies within R_{180}. Around these 268 clusters, we select galaxies in a slightly wider redshift range, lying within $2\,R_{100} = 2.6\,R_{180}$ from the cluster in projected space, with absolute line-of-sight velocity relative to the cluster mean (ALOSV) within $3\sigma_\upsilon$, with absolute magnitude $M_r < -20.5$, and with angular effective radius $\theta_{\rm eff} \le 5''$ to avoid large galaxies for which the SDSS fibers only span the inner region. This selection yields 19 904 galaxies and we also select 21,000 field galaxies that lie at least 10 Mpc in projection away from any cluster (corresponding to typically 6.7 $R_{\rm v}$, where the virial radius $R_{\rm v} \equiv R_{100}$).

We identify GORES as galaxies with both $H_\delta > 2$ Å and $D_n4000 < 1.5$ (age less than 1 Gyr for most galaxies, [9]). The symbols in the left panel of Fig. 1 illustrate how the observed fraction of GORES increases with projected radius. More important, this plot shows that this radial segregation of GORES fraction is modulated by ALOSV: there are significantly more (fewer) GORES among the higher (lower) ALOSV galaxies.

3 Deprojection

We first deproject the fraction of GORES without taking into account their ALOSVs. Writing $g_{\text{GORES}}(R) = N_{\text{GORES}}(R)/N_{\text{tot}}(R)$, we can deduce the surface densities, $\Sigma(R) = N(R)/(2\pi R)$ of GORES and of all galaxies, and then perform Abel deprojection yielding space densities $\nu(r) = -(1/\pi)\int_R^\infty \Sigma'(R)\,dR/\sqrt{R^2 - r^2}$ for both GORES and all galaxies. The right panel of Fig. 1 shows that this deprojection leads to a fraction of GORES that rises linearly with physical radius and then saturates as

$$f_{\text{GORES}}(r) = f_0 \frac{r}{r+a}, \tag{1}$$

with best-fit parameters of $f_0 = 0.52$ and $a = 1.26\,R_{\text{v}}$.

We now assume that the observed fraction of GORES in projected phase space, $g_{\text{GORES}}(R, |v_{\text{LOS}}|)$, is also a fraction of the dynamical *class* of galaxies: we distinguish virialized, infalling and backsplash galaxies according to their $z = 0$ position in radial phase space in a cosmological simulation as shown in the left panel of Fig. 2. The right panel of Fig. 2 shows that in our standard scheme, where particles in regions A and B are both infalling (this scheme provides the best fits to the observed fraction of GORES in projected phase space), the fraction of backsplash particles reaches over 50% just beyond the virial radius and for the lowest ALOSVs.

We suppose that the fraction of GORES varies with physical radius r as in (1), varying the normalization and possibly the scale for each class:

$$f_\alpha(r) = f_\alpha \frac{r/R_{\text{v}}}{r/R_{\text{v}} + a_\alpha}, \tag{2}$$

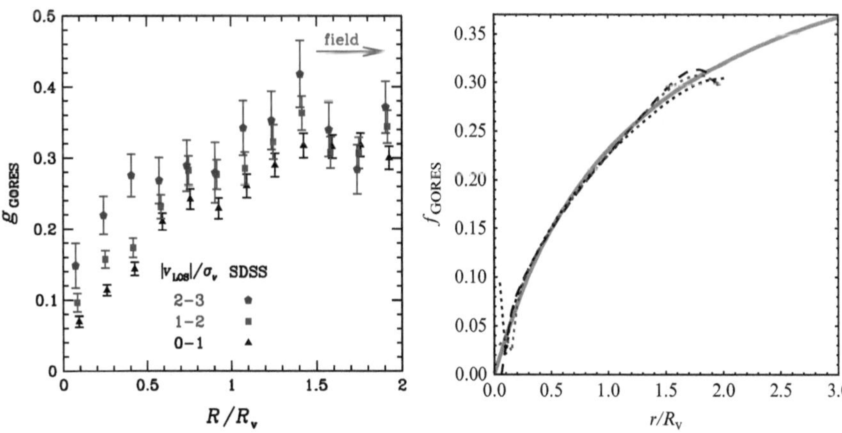

Fig. 1 *Left*: Observed fraction of GORES in three bins of ALOSV. *Right*: Deprojected fractions of GORES. *Thin curves* show polynomial fits of the log surface density vs. log R (orders 3, 4, and 5, *dotted*, *dashed*, and *dash-dotted*, respectively). The *longer thick curve* is (1) with $f_0 = 0.52$ and $a = 1.26\,R_{\text{v}}$, obtained by a χ^2 fit to the order 4 polynomial fit

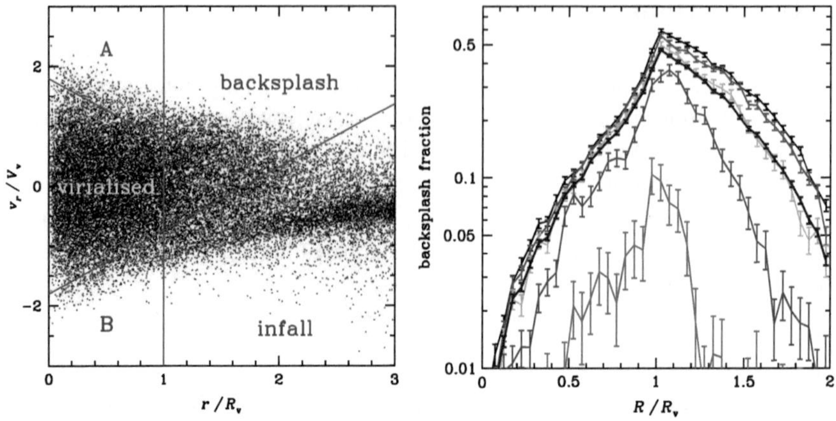

Fig. 2 *Left*: Radial phase space distribution (in virial units) of dark matter particles of a stack of 93 mock regular mock clusters from a cosmological hydrodynamical simulation [6]. The critical velocity separating infall from backsplash population (*long diagonal line*) is taken from a single halo of an older, dissipationless, cosmological simulation [15]. For clarity, only 1 out of 550 particles of the original simulation is plotted. The *letters* indicate uncertain classes. *Right*: Fraction of backsplash particles in projected phase space, for our standard scheme, particles in regions A and B are infalling. The *thin lines* indicate ALOSV bins increasing from 0.25, 0.75,...,1.75, and > $2\,\sigma_\upsilon$, going from top to bottom, while the *thick line* shows the average

that is rising roughly linearly with radius for $r < a_\alpha$, saturating to an asymptotic value f_α at large radii. The predicted fractions of GORES is then

$$g_{\rm GORES}(R_i, \upsilon_j) = \sum_\alpha p(\alpha|R_i, \upsilon_j) \sum_k f_\alpha(r_k)\, q(r_k|R_i, \upsilon_j, \alpha), \qquad (3)$$

where $p(\alpha|R_i, \upsilon_j)$ is the probability that a particle located at projected radius R_i and ALOSV υ_j is of class α, while $q(k|R_i, \upsilon_j, \alpha)$ is the fraction of the particles of class α in the cell of projected phase space (R_i, υ_j) that are in the kth bin of physical radius (with $r_k = R_i \cosh u_k$, where u_k is linearly spaced from 0 to $\cosh^{-1} r_{\rm max}/R_i$, using $r_{\rm max} = 50\,R_{\rm v}$). We measure p and q from the dark matter particles in the stack of 93 regular mock clusters [11] in a hydrodynamical cosmological simulation [6].

Figure 3 shows how the best-fitting models for the two best fitting schemes (right panel, both with B = infall, but with A = infall and A = backsplash, respectively leading to $\chi^2 = 1.2$ and 1.4 per degree of freedom with cosmic variance from the simulation included) fit the observed fraction of GORES in projection (left panel). Schemes where regions A and B are both virialized or both backsplash, or where $f_{\rm GORES}$ is constant per class produce substantially worse fits.

Interestingly, the two best-fitting models yield a GORES fraction that is independent of radius for the infall population. Moreover, in the best fitting scheme, the galaxies bouncing out of the cluster with very high positive line-of-sight velocities are of the infalling class. This suggests that the quenching of efficient star formation

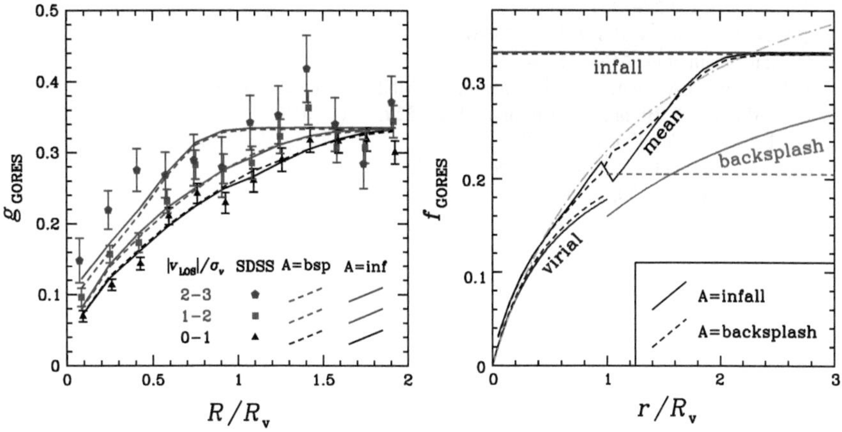

Fig. 3 *Left*: Observed fraction of GORES with two best-fitting models ((2) using (3)) overplotted, both assume that the particles in B are of the infalling class. *Right*: Deprojected fraction of GORES for the two best-fitting models. The *dash-dotted curve* is the global fit of (1)

is not instantaneous as the infalling galaxies pass through the pericenter of their orbit. Moreover, inspection of the right panel of Fig. 3 indicates that at $r = R_v$, where the three galaxy classes can be compared, the backsplash GORES fraction is much closer to that of the virialized class than to the infall GORES fraction (in fact in our best fitting scheme A = infall, the backsplash fraction of GORES is lower than the corresponding fraction for the virial class, but forcing equal fractions of GORES for the two classes at the virial radius produces an equally good fit). This suggests that as galaxies cross through clusters, efficient star formation is nearly completely quenched once galaxies reach the virial sphere, either after the first pericenter on their way out, or at least before they are mixed with the virialized galaxies.

While GORES account for 18% of the galaxies within the virial cone, in our best fitting models, they also account for 13% within the virial sphere, 34% and 19% of the infalling and backsplash classes, respectively. GORES also account for as many as 11% of the virialized galaxies, perhaps the consequence of star formation resulting from tides caused by interactions with other galaxies.

References

1. C. Adami, A. Biviano, A. Mazure, A&A **331**, 439 (1998)
2. M.L. Balogh et al., ApJL **615**, L101 (2004)
3. M.L. Balogh et al., MNRAS **348**, 1355 (2004)
4. A. Biviano et al., ApJ **396**, 35 (1992)
5. A. Biviano et al., A&A **321**, 84 (1997)
6. S. Borgani et al., MNRAS **348**, 1078 (2004)

7. A. Dressler, ApJ **236**, 351 (1980)
8. C.P. Haines et al., ApJl **647**, L21 (2006)
9. G. Kauffmann et al., MNRAS **341**, 33 (2003)
10. S. Mahajan, G.A. Mamon, S. Raychaudhury, MNRAS, in press, arXiv:1106:3062
11. G.A. Mamon, A. Biviano, G. Murante, A&A **520**, A30 (2010)
12. C. Moss, MNRAS **373**, 167 (2006)
13. C. Moss, R.J. Dickens, MNRAS **178**, 701 (1977)
14. K. Rines et al., AJ **130**, 1482 (2005)
15. T. Sanchis, E. Łokas, G.A. Mamon, A&A **414**, 445 (2004)
16. X. Yang et al., ApJ **671**, 153 (2007)

Simulations of Star Counts and Galaxies Towards Vista Variables in the via Láctea Survey Region

E. Amôres, N. Padilla, L. Sodré, D. Minniti, and B. Barbuy

Abstract In the present work, we present results of simulations for star and galaxy counts using a Galaxy and Large-Scale Structure Model (respectively) towards the VVV survey region. In total there are around 3 billion and 2 million simulated stars and galaxies, respectively, up to the survey limiting magnitude. The simulations were performed using the most recent and realistic extinction models. In the case of the galaxy simulations, these come from the semi-analytic galaxy formation model of [Bower et al., MNRAS 370, 645 (2006)] applied to the Millennium simulation [Springel et al., Nature 435, 629 (2005)]. Our results consist of color-color and color-magnitude diagrams in the space of colors (JHKYZ) for stars and galaxies. They also include the expected observed distribution of background galaxies, as well as the expected number of stars that will be observed by VVV in any range of magnitude, color, spectral type and distance for each galactic component (disk, bulge and halo), which is also useful for several VVV projects.

1 Star Counts Model

In order to predict star counts distribution towards the VVV region we made use of TRI–LEGAL Galaxy Model [2] with galactic parameters as pointed by [3] and [2] considering the VVV completeness limit, i.e. K = 20.0. For the

E. Amôres (✉)
SIM, Faculdade de Ciências da Universidade de Lisboa, Lisbon, Portugal
e-mail: amores@sim.ul.pt

N. Padilla
Departamento de Astronomía y Astrofísica, PUC, Chile

L. Sodré · B. Barbuy
IAG – USP, Brazil

D. Minniti
Departamento de Astronomía y Astrofísica, PUC and Vatican Observatory, Chile

I. Ferreras and A. Pasquali (eds.) *Environment and the Formation of Galaxies: 30 years later*, Astrophysics and Space Science Proceedings,
DOI 10.1007/978-3-642-20285-8_28,

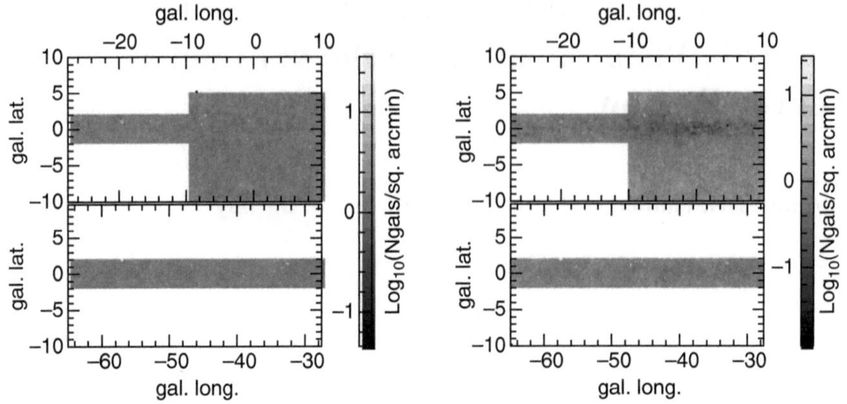

Fig. 1 Galaxy counts without (*left*) and with (*right*) MW extinction

other filters, we have cut (after performing the simulation) at other VVV limits, i.e. $Z = 21.6, Y = 20.9, J = 20.6, H = 19.0$. We have used the [4] interstellar extinction model. We estimate a total of 9.2 billion of stars towards the VVV region.

2 Galaxy Counts

With the aim of producing a simulation of the background galaxy counts, we used semi-analytic galaxies from [1] applied to the Millennium simulation [5], which includes both intrinsic extinction for each galaxy as well as extinction due to the MW ISM. The results with and without the MW extinction can be seen in the left and right panels of Fig. 1 (respectively), where for a fixed magnitude limit, a clear drop in the number counts of galaxies in the disk and bulge of the MW can be seen (left). This changes the counts of galaxies as a function of redshift by shifting the distribution to lower distances (by $\Delta z = -0.05$), and removing about 10% of the galaxies in the sample with no MW extinction. For instance, considering 2MASS extended sources [6] there are 342 objects classified as galaxies by 2MASS towards the full VVV region which gives a density of 0.65 galaxies per square degree.

Acknowledgements The VVV Survey is supported by the European Southern Observatory, BASAL Center for Astrophysics and Associated Technologies PFB-06, FONDAP Center for Astrophysics 15010003, and MIDEPLAN Milky Way Millennium Nucleus P07-021-F. Some VVV tiles were made using Aladin sky atlas, SExtractor software and products from TERAPIX pipeline. Eduardo Amôres obtained financial support for this work provided by and Fundação para a Ciência e Tecnologia (FCT) under the grant SFRH/BPD/42239/2007. Laerte Sodré Jr. and Beatriz Barbuy achnowledges the support of the support of FAPESP and CNPq.

References

1. R. Bower et al., MNRAS **370**, 645 (2006)
2. L. Girardi et al., A&A **436**, 895 (2005)
3. E. Vanhollebeke, M.A.T. Groenewegen, L. Girardi, MNRAS **498**, 95 (2009)
4. D.J. Marshall et al., A&A **453**, 635 (2006)
5. V. Springel et al., Nature **435**, 629 (2005)
6. M.F. Skrutskie et al., AJ **133**, 1163 (2006)

Nature and Nurture of Early-Type Dwarf Galaxies in Low Density Environments

R. Grützbauch, F. Annibali, R. Rampazzo, A. Bressan, and W.W. Zeilinger

Abstract We study stellar population parameters of a sample of 13 dwarf galaxies located in poor groups of galaxies using high resolution spectra observed with VIMOS at the ESO-VLT [Grützbauch et al., A&A **502**, 473 (2009)]. LICK-indices were compared with Simple Stellar Population models to derive ages, metallicities and [α/Fe]-ratios. Comparing the dwarfs with a sample of giant ETGs residing in comparable environments we find that the dwarfs are on average younger, less metal-rich, and less enhanced in alpha-elements than giants. Age, Z, and [α/Fe] ratios are found to correlate both with velocity dispersion and with morphology. We also find possible evidence that low density environment (LDE) dwarfs experienced more prolonged star formation histories than Coma dwarfs, however, larger samples are needed to draw firm conclusions.

1 Age, Z, and [α/Fe] as a Function of Velocity Dispersion, Morphology and Environment

We confirm that the correlations between age, metallicity (Z), [α/Fe] and velocity dispersion (σ) found for giant galaxies [1] extend towards the low mass regime of our dwarf sample, albeit with a larger scatter (Fig. 1, left panel). The relatively tight Z–σ relation, however, suggests that the scatter in age and [α/Fe] is intrinsic, reflecting the variety of star formation histories in dwarf galaxies.

R. Grützbauch (✉)
University of Nottingham, Nottingham, UK
e-mail: ruth.grutzbauch@nottingham.ac.uk

F. Annibali · R. Rampazzo · A. Bressan
INAF Osservatorio Astronomico di Padova, Padova, Italy

W.W. Zeilinger
Department of Astronomy, University of Vienna, Vienna, Austria

I. Ferreras and A. Pasquali (eds.) *Environment and the Formation of Galaxies: 30 years later*, Astrophysics and Space Science Proceedings,
DOI 10.1007/978-3-642-20285-8_29, © Springer-Verlag Berlin Heidelberg 2011

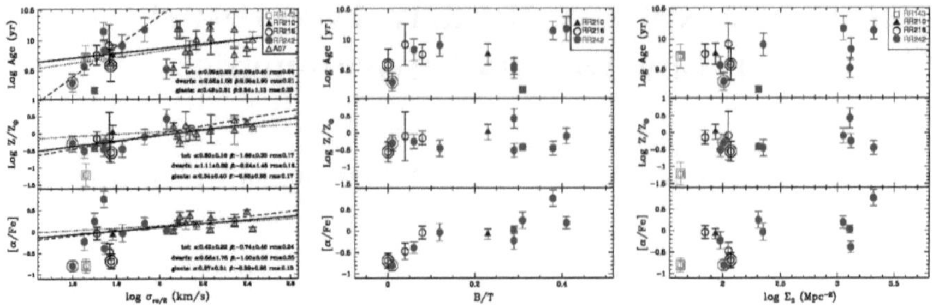

Fig. 1 Age, metallicity and [α/Fe] as a function of velocity dispersion σ (*left*) bulge fraction B/T (*mid*) and local density Σ_3 (*right*). *Triangles* are massive ETGs from [1]

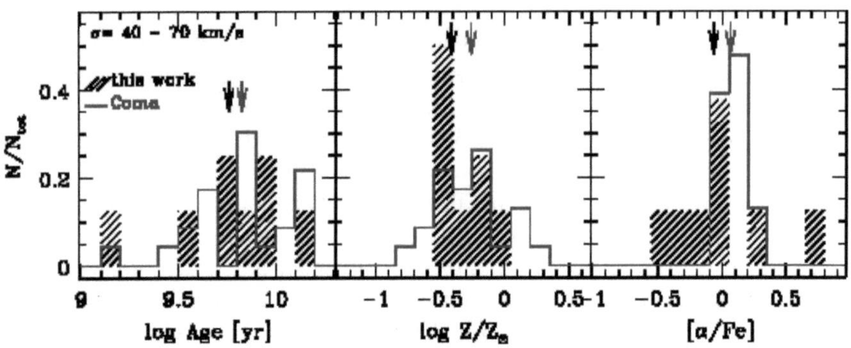

Fig. 2 Distribution of age, metallicity and [α/Fe] in poor groups (*shaded*) compared to Coma (*grey*)

We find a strong correlation between the bulge fraction (B/T) and the [α/Fe]-ratio, such that galaxies with a stronger bulge have shorter SF timescales (Fig. 1, mid panel). The same correlation is found for the Sersic index n. The presence of a morphology-[α/Fe] relation seems in contradiction to the possible evolution along the Hubble sequence from low to high B/T galaxies. In other words, systems similar to present-day dwarf galaxies could not be the building blocks of today's giant ellipticals.

Local densities are measured for each galaxy based on the distance to the third nearest neighbour. We find a weak correlation between local density (Σ_3) and [α/Fe] for the dwarf sample, whereas age and metallicity are not clearly correlated with Σ_3 (Fig. 1, right panel). Note however, that Σ_3 is a projected density and projection effects can considerably distort the intrinsic 3D galaxy density.

Finally, we compare our results to red, passive dwarf galaxies in the Coma Cluster studied by [3] (see Fig. 2). Statistical tests indicate that the only significant difference between our sample and the Coma sample is in the distribution of [α/Fe], with LDE dwarfs having experienced a more prolonged star formation than cluster dwarfs. By contrast, no strong difference in the [α/Fe] ratios is observed for giant ETGs in the field and in the cluster.

References

1. F. Annibali et al., A&A **463**, 455 (2007)
2. R. Grützbauch et al., A&A **502**, 473 (2009)
3. R.J. Smith et al. MNRAS **392**, 1265 (2009)

Galaxy Evolution in Clusters Since $z \sim 1$

A. Aragón-Salamanca

Abstract Galaxy clusters provide some of the most extreme environments in which galaxies evolve, making them excellent laboratories to study the age old question of "nature" vs. "nurture" in galaxy evolution. Here I review some of the key observational results obtained during the last decade on the evolution of the morphology, structure, dynamics, star-formation history and stellar populations of cluster galaxies since the time when the Universe was half its present age. Many of the results presented here have been obtained within the ESO Distant Cluster Survey (EDisCS) and Space Telescope A901/02 Galaxy Evolution Survey (STAGES) collaborations.

1 Introduction

The precise role that environment plays in shaping galaxy evolution is still hotly debated. Trends to passive and/or more spheroidal populations in dense environments are widely observed: galaxy morphology [11, 12], colour [2, 6, 29], star-formation [17, 31], and stellar age and AGN fraction [28] all correlate with measurements of the local galaxy density. Disentangling the relative importance of internal and external physical mechanisms responsible for these relations is challenging. It is natural to expect that high-density environments will preferentially host older stellar populations. Hierarchical models of galaxy formation (e.g. [8]) suggest that galaxies in the highest density peaks started forming stars and assembling mass earlier. Simultaneously, galaxies forming in high-density environments will have more time to experience the external influence of their local environment. Those processes will also act on infalling galaxies as they are continuously accreted

A. Aragón-Salamanca (✉)
School of Physics and Astronomy, University of Nottingham, Nottingham NG7 2RD, UK
e-mail: alfonso.aragon@nottingham.ac.uk

I. Ferreras and A. Pasquali (eds.) *Environment and the Formation of Galaxies: 30 years later*, Astrophysics and Space Science Proceedings,
DOI 10.1007/978-3-642-20285-8_30,

into larger haloes. There are many plausible physical mechanisms by which a galaxy could be transformed by its environment: removal of the hot or cold gas supply through ram-pressure stripping [20, 30]; tidal effects leading to halo truncation [5] or triggered star formation through gas compression [15]; interactions between galaxies themselves via low-speed major mergers [4] or frequent impulsive encounters ("harassment" [35]). Galaxy clusters are excellent laboratories to study all these processes since they provide some of the most extreme environments in which galaxies evolve.

In addition to all these environmental effects, luminosity (or more directly, mass) is also critical in regulating how susceptible a galaxy is to external influences. For example, in low density environments the fraction of passive galaxies is a strong function of luminosity [21]. Understanding these issues in full is further complicated by the relative amounts of obscured and unobscured star formation that may or may not be present. Moreover, changes in morphology are not necessarily equivalent to changes in star formation. There is no guarantee that external processes causing an increase or decrease in the star-formation rate act on the same timescale, to the same degree, or in the same regime as those responsible for structural changes.

Despite all these complications, significant progress has been made. In this contribution I review some of the key observational results obtained during the last decade or so on the evolution of the morphology, structure, dynamics, star-formation history and stellar populations of cluster galaxies since the time when the Universe was half its present age. Most of the results presented here have been obtained within the ESO Distant Cluster Survey (EDisCS) and Space Telescope A901/02 Galaxy Evolution Survey (STAGES) collaborations. I thank the members of these collaborations for allowing me to quote results obtained with their help.

2 Results from the ESO Distant Cluster Survey

The ESO Distant Cluster Survey (EDisCS) [40, 42, 48] is a multiwavelength survey of galaxies in 20 fields containing galaxy clusters at $z = 0.4$–1. Because they were optically selected from the Las Campanas Distant Cluster Survey [18] they span a very broad range of velocity dispersions (and masses), making this the ideal sample to study the high-redshift counterparts of present-day clusters [34]. The available data for these clusters include deep optical and near-IR photometry [48], deep multi-slit spectroscopy [22, 34], and ACS/HST mosaic imaging of 10 of the highest redshift clusters [10]. Photometric redshifts have been derived from the optical and near-IR imaging [36]. Additional follow-up includes XMM-Newton X-Ray observations [27], Spitzer IRAC and MIPS imaging [14], and Hα narrow-band imaging [13]. Our main results are:

1. The colour and rest-frame K-band luminosity evolution since $z \sim 1$ of the brightest cluster galaxies (BCGs) are in good agreement with population synthesis models of stellar populations which formed at $z > 2$ and evolved passively

thereafter. In contrast with some previous results [1], we do not detect any significant change in the stellar mass of the BCGs. These results do not seem to depend on the velocity dispersion of the parent cluster. However, we do find a correlation between the velocity dispersion of the clusters and the stellar mass of the BCGs in the sense that the clusters with large velocity dispersions/masses tend to have more massive BCGs. This dependency, although significant, is relatively weak: the stellar mass of the BCGs changes only by ~70% over a two order of magnitude range in cluster mass. This dependency does not change significantly with redshift [47].

2. While the colours of red-sequence galaxies are well described by an old, passively evolving population, there is a significant deficit of faint red galaxies in EDisCS clusters compared with their present-day counterparts. This decrease in the luminous-to-faint ratio of red galaxies since $z \sim 0.8$ is in qualitative agreement with predictions of a model where the blue bright galaxies that populate the colour-magnitude diagram of high-redshift clusters, have their star formation suppressed by the hostile cluster environment. This indicates that the red-sequence population of high-redshift clusters does not contain all progenitors of nearby red-sequence cluster galaxies. A significant fraction of these must have moved on to the red sequence below $z \sim 0.8$ [7, 9, 41].

3. The intrinsic colour scatter about the colour-magnitude relation (CMR) for morphologically-selected E and S0 galaxies in all EDisCS clusters is very small. This implies that by the time cluster elliptical and S0 galaxies achieve their morphology, the vast majority have already joined the red sequence. However, there is a small minority of faint early-type galaxies (7%) that are significantly bluer than the CMR. If the colour scatter is due to differences in stellar population ages, the formation redshift z_F of the E and S0 galaxies does not depend on the cluster velocity dispersion. However, z_F increases weakly with cluster redshift, suggesting that, at any given redshift, in order to have a population of fully formed ellipticals and S0s they needed to have formed most of their stars $\gtrsim$ 2–4 Gyr prior to observation. That does not mean that all early-type galaxies in all clusters formed at these high redshifts. It means that the ones we see already having early-type morphologies also have reasonably old stellar populations. This is partly a manifestation of the 'progenitor bias', but also a consequence of the fact that the vast majority of the early-type galaxies in clusters (in particular the massive ones) already had old stellar populations by the time they achieved their morphology. Elliptical and S0 galaxies exhibit very similar colour scatter, implying similar stellar population ages. The scarcity of blue S0s indicates that, if they are the descendants of spirals whose star formation has ceased, the parent galaxies were already red when they became S0s. This suggests the red spirals found preferentially in dense environments [3] could be the progenitors of cluster S0s. We also find that fainter early-type galaxies finished forming their stars later, consistent with the cluster red sequence being built over time and the brightest galaxies reaching the red sequence earlier than fainter ones. Finally, combining the CMR scatter analysis with the observed evolution in the CMR zero-point we find that the early-type cluster galaxy population must have had their star formation truncated/stopped over an extended period $\Delta t \gtrsim 1$ Gyr [26].

4. High S/N VLT spectra reveal that the evolution of the stellar population properties of red-sequence galaxies depend on their mass: while the properties of the most massive are well described by passive evolution and high-redshift formation, those of the less massive galaxies are consistent with a more extended star-formation history [44]. The evolution of the absorption line strengths for the red-sequence galaxies can be reproduced if 40% of the galaxies with $\sigma < 175\,\mathrm{km\,s^{-1}}$ entered the red-sequence between $z = 0.75$ and $z = 0.45$, in excellent agreement with the evolution of the luminosity functions [7, 9, 41]. Moreover, the fraction of red-sequence galaxies exhibiting early-type morphologies (E and S0) decreases by 20% from $z = 0.75$ to $z = 0.45$. This can be understood if the red-sequence becomes more populated at later times with disc galaxies whose star formation has been quenched. Thus, the processes quenching star formation do not necessarily produce a simultaneous morphological transformation of the galaxies entering the red-sequence [44].

5. We have also studied the evolution of the fundamental plane (FP) for galaxies in clusters with a very broad range of masses. If we interpret the evolution in the FP as due only to changes in the luminosity of the galaxies, the mass-to-light ratio would evolve as $\Delta\log(M/L_{\mathrm{B}}) = (-0.54 \pm 0.01)z = (-1.61 \pm 0.01)\log(1+z)$ for cluster galaxies and as $\Delta\log(M/L_{\mathrm{B}}) = (-0.76 \pm 0.01)z = (-2.27 \pm 0.03)$ $\log(1+z)$ in the field. However, that cannot be the whole story since the galaxy sizes and velocity dispersions also evolve. Taken together, the variations in size and velocity dispersion imply that the luminosity evolution with redshift derived from the zero point of the FP is somewhat milder than that derived without taking these variations into account. At fixed dynamical masses, the effects of size and velocity dispersion variations almost cancel out. At fixed stellar masses, the luminosity evolution is reduced to $L_B \propto (1+z)^{1.0}$ for cluster galaxies and $L_B \propto (1+z)^{1.67}$ for field galaxies. This luminosity evolution implies that massive ($> 10^{11}\,\mathrm{M_{\odot}}$) cluster galaxies are old ($z_{\mathrm{F}} > 1.5$) and lower mass galaxies are ~3–4 Gyr younger, confirming the picture of a progressive build-up of the red sequence in clusters. Field galaxies follow the same trend, but are ~1 Gyr younger at a given redshift and mass [43].

6. Combining previously published data with ACS-based visual morphologies for 10 $0.5 < z < 0.8$ EDisCS clusters we find no significant evolution in the mean fractions of elliptical, S0, and late-type (Sp+Irr) galaxies in clusters over the redshift range $0.5 < z < 1.2$. In contrast, existing studies of lower redshift clusters have revealed a factor of ~2 increase in the typical S0 fraction between $z = 0.4$ and 0, accompanied by a commensurate decrease in the Sp+Irr fraction and no evolution in the elliptical fraction. It seems therefore that cluster morphological fractions plateau beyond $z \sim 0.4$ [10]. The evolution in the morphological fractions is accompanied by a parallel evolution in the corresponding stellar mass fractions for the different morphological classes [46]. There seems to be a mild correlation between morphological content and cluster velocity dispersion, highlighting the importance of careful sample selection in evaluating evolution [10]. However, alternative analysis using the galaxies' structural parameters as proxies for morphology do not find such a trend [45].

7. We have also explored how the proportion of star-forming galaxies evolves between $z = 0.8$ and 0 as a function of galaxy environment. At high z most systems follow a broad anticorrelation between the fraction of star-forming galaxies and the cluster velocity dispersion $\sigma_{\rm clus}$. At face value, this suggests that at $z = 0.4$–0.8 the mass of the system largely determines the proportion of galaxies with ongoing star formation. At these redshifts the strength of star formation in star-forming galaxies is also found to vary systematically with environment. In contrast, local SDSS clusters have much lower fractions of star-forming galaxies than clusters at $z = 0.4$–0.8 and show a plateau for $\sigma_{\rm clus} \geq 550\,{\rm km\,s^{-1}}$, where the fraction of star forming galaxies does not vary systematically with $\sigma_{\rm clus}$. This means that the fraction of star-forming galaxies evolves more strongly in intermediate-mass systems ($\sigma_{\rm clus} \sim 500$–$600\,{\rm km\,s^{-1}}$ at $z \sim 0$) [37].

8. We find that the star-formation properties and morphologies of galaxies in clusters and groups at $z = 0.4$–0.8 depend on projected local galaxy density. In both nearby (SDSS) and distant clusters, higher density regions contain proportionally fewer star-forming galaxies, but the average [OII] equivalent width of star-forming galaxies is independent of local density. In distant clusters the average current star formation rate (SFR) in star-forming galaxies seems to peak at intermediate densities. Unlike at low-z, at high-z the relation between star-forming fraction and local density seems to depend on the cluster mass. The morphology-density relation is already well established in $z = 0.4$–0.8 clusters, but is completely dominated by radial variations in the spiral and elliptical fractions, with the S0s playing no role given their small numbers. The decline of the spiral fraction with density is entirely driven by galaxies of type Sc or later. For galaxies of a given Hubble type, we see no evidence that their star formation properties depend on local environment [38].

9. The incidence of post-starburst (E+A or k+a) galaxies, where the star formation ended abruptly in the past Gyr, depends strongly on environment at intermediate redshifts. They reside preferentially in clusters and, unexpectedly, in a subset of the $\sigma_{\rm clus} = 200$–$400\,{\rm km\,s^{-1}}$ groups, those that have a low fraction of [OII] emitters. In these environments, 20–30% of the star-forming galaxies have had their star formation activity recently truncated. In contrast, there are proportionally fewer k + a galaxies in the field, the poor groups, and groups with a high [OII] fraction. Moreover, the incidence of k+a galaxies correlates with the cluster velocity dispersion: more massive clusters have higher proportions of k+a's. Dusty starbursts present a very different environmental dependence. They are numerous in all environments at $z = 0.4$–0.8, but they are especially numerous in all groups, favouring the hypothesis that dusty starbursts are triggering by mergers. We conclude that cluster k+a galaxies are mainly rather massive S0 and Sa galaxies observed in a transition phase, evolving from star-forming, recently-infallen later types to passively-evolving cluster early-type galaxies. The correlation between k+a fraction and cluster velocity dispersion supports the hypothesis that in clusters these galaxies originate from processes related to the intracluster medium [39].

10. Using spatially-resolved VLT spectra of EDisCS galaxies with emission lines we find that disturbances in the galaxies' gas kinematics are more frequent in clusters than in the field. This effect is stronger for brighter galaxies. This suggests that

gas is being stripped off very efficiently from faint cluster galaxies, removing them from the emission-line sample. Moreover, the fraction of kinematically-disturbed galaxies increases with cluster velocity dispersion and decreases with distance from the cluster centre, but it remains constant with projected galaxy density. This strongly suggest that the effect must be caused by ICM-related processes. The fraction of morphologically-disturbed galaxies (from HST/F814W images) does not depend on luminosity or environment. For the kinematically-undisturbed galaxies, we find that the cluster and field Tully-Fisher relations are remarkably similar. However, we find some tentative evidence that the specific star-formation rate (SFR per unit stellar mass) is suppressed in the kinematically disturbed galaxies. We also find that cluster galaxies show truncated gas discs as compared with similarly-selected field galaxies. These results suggest that environmental effects are mild enough not to disturb the stellar discs while being able to strongly affect the gas in cluster galaxies [25, 26].

3 Results from the Space Telescope A901/2 Galaxy Evolution Survey

The Space Telescope A901/2 Galaxy Evolution Survey (STAGES) is a multiwavelength project designed to probe physical drivers of galaxy evolution across a wide range of environments and luminosity [19]. This project, targeting the Abell 901(a,b)/902 multiple cluster system (hearafter A901/2) at $z \sim 0.165$, allows us to carry out a comprehensive study of the environmental processes at work in shaping galaxy evolution. The survey design addresses simultaneously several key areas: a wide range of environments; a wide range in galaxy luminosity; and sensitivity to both obscured and unobscured star formation, stellar masses, AGN, and detailed morphologies. Furthermore, STAGES provides several complementary measurements of 'environment' in order to understand directly the relative influences of the local galaxy density, the hot ICM and the dark matter on galaxy transformation. A further advantage is given by examining systems that are not simply massive clusters already in equilibrium. By including systems in the process of formation (when extensive mixing has not yet erased the memory of early timescales), the various environmental proxies listed above might still be disentangled.

The complex multi-cluster A901/2 system at $z \sim 0.165$ has been the subject of an 80-orbit F606W HST/ACS mosaic covering the full $0.5° \times 0.5°$ ($\sim 5 \times 5\,\mathrm{Mpc}^2$) span of the system. Extensive multiwavelength observations with XMM-Newton, GALEX, Spitzer, 2dF, GMRT, Magellan, and the 17-band COMBO-17 photometric redshift survey complement the HST imaging. Our survey goals include simultaneously linking galaxy morphology with other observables such as age, star-formation rate, nuclear activity, and stellar mass. In addition, with the multiwavelength dataset and new high resolution mass maps from gravitational lensing [24], we are able to disentangle the large-scale structure of the system. By examining all aspects of environment we are evaluating the relative importance of the dark matter halos, the local galaxy density, and the hot X-ray gas in driving galaxy transformation.

In summary, STAGES focuses on a single large-scale structure which samples a very broad range of galaxy environments and masses at a single redshift, allowing us to decouple environmental and stellar mass effects from redshift-related ones. I will summarise below some of the results more relevant to this symposium.[1]

1. We have explored the amount of obscured star formation as a function of environment by combining the UV/optical SEDs from COMBO-17 with Spitzer 24 μm photometry. We find that while there is an overall suppression in the fraction of star-forming galaxies with density, the small amount of star formation surviving the cluster environment is to a large extent obscured [16]. Furthermore, we also find that passive spiral and dusty red galaxies in the cluster system are largely the same population. These galaxies form stars at a substantial rate that is only a factor of ~4 times lower than blue spirals at fixed mass, but their star formation is more obscured and has weak optical signatures. They constitute over half of the star forming galaxies at masses above $\log M_*/M_\odot = 10$ and are thus a vital ingredient for understanding the overall picture of star-formation quenching in cluster environments [49].

2. Only $4.9 \pm 1.3\%$ of intermediate mass ($M_* \geq 1 \times 10^9 M_\odot$) galaxies in the STAGES field are interacting. The interacting galaxies are found to lie outside the cluster cores and to be concentrated in the region between the cores and virial radii of the clusters. The average star formation rate is enhanced only by a modest factor in interacting galaxies compared to non-interacting galaxies. Interacting galaxies only contribute ~20% of the total SFR density in the A901/2 clusters [23].

3. We explored the dependence of the mass-size-relation with environment using a large sample of morphologically-classified galaxies. For elliptical, lenticular and high-mass ($\log M_*/M_\odot > 10$) spiral galaxies we find no evidence to suggest any such environmental dependence, implying that internal drivers are governing their size evolution. For intermediate-/low-mass spirals ($\log M_*/M_\odot < 10$) we find tentative evidence for a possible environmental dependence: the mean effective radius for lower mass spirals is ~15–20% larger in the field than in the cluster. This is due to a population of low-mass field spirals with relatively extended disks that are largely absent from the cluster environments. These galaxies contain extended stellar discs not present in their cluster counterparts. This suggests that the fragile extended stellar discs of these spiral galaxies may not survive the environmental conditions in the cluster. Internal physical processes seem to be the main drivers governing the size evolution of galaxies, with the environment possibly playing a role affecting only the discs of low-mass spirals [32]. Interestingly, a detailed study of the spiral galaxy population using STAGES ACS images does not find any environmental effects on the detailed structure of their stellar disks such as their

[1]For a more complete and up-to-date description of STAGES-related science see http://www.nottingham.ac.uk/astronomy/stages/.

scale lengths or the presence or absence of disk truncations/anti-truncations. This implies that the galaxy environment is not affecting the stellar distribution in the outer stellar disk [33].

4 Conclusions

The results quoted above allow us to conclude, among many other things, that spirals transform into S0s in clusters, with the most massive ones completing the process earlier. Star formation ceases quickly in in-falling spirals, leaving a "k+a" signature. Infalling blue spirals become (dusty) red spirals and then S0s. The truncation of star formation proceeds outside-in, helping to grow bulges. The driving mechanism is "gentle" since it preserves disks. All these processes happen in massive clusters first.

References

1. A. Aragón-Salamanca, C.M. Baugh, G. Kauffmann, MNRAS **297**, 427 (1998)
2. I.K. Baldry et al., MNRAS **373**, 469 (2006)
3. S.P. Bamford et al., MNRAS **393**, 1324 (2009)
4. J.E. Barnes, ApJ **393**, 484 (1992)
5. K. Bekki, ApJL **510**, L15 (1999)
6. M.R. Blanton et al., ApJ **629**, 143 (2005)
7. G. De Lucia et al., ApJL **610**, L77 (2004)
8. G. De Lucia et al., MNRAS **366**, 499 (2006)
9. G. De Lucia et al., MNRAS **374**, 809 (2007)
10. V. Desai et al., ApJ **660**, 1151 (2007)
11. A. Dressler, ApJ **236**, 351 (1980)
12. A. Dressler et al., ApJ **490**, 577 (1997)
13. R.A. Finn et al., ApJ **630**, 206 (2005)
14. R.A. Finn et al., ApJ **720**, 87 (2010)
15. Y. Fujita, ApJ **509**, 587 (1998)
16. A. Gallazzi et al., ApJ **690**, 1883 (2009)
17. P.L. Gómez et al., ApJ **584**, 210 (2003)
18. A.H. Gonzalez et al., ApJS **137**, 117 (2001)
19. M.E. Gray et al., MNRAS **393**, 1275 (2009)
20. J.E. Gunn, J.R. Gott III, ApJ **176**, 1 (1972)
21. C.P. Haines et al., ApJL **647**, L21 (2006)
22. C. Halliday et al., A&A **427**, 397 (2004)
23. A. Heiderman et al., ApJ **705**, 1433 (2009)
24. C. Heymans et al., MNRAS **385**, 1431 (2008)
25. Y.L. Jaffé, this volume (2010). arXiv:1011.6525
26. Y.L. Jaffé et al., MNRAS **410**, 280 (2011)
27. O. Johnson et al., MNRAS **371**, 1777 (2006)
28. G. Kauffmann et al., MNRAS **353**, 713 (2004)
29. T. Kodama et al., ApJL **562**, L9 (2001)
30. R.B. Larson, B.M. Tinsley, C.N. Caldwell, ApJ **237**, 692 (1980)
31. I. Lewis et al., MNRAS **334**, 673 (2002)

32. D.T. Maltby et al., MNRAS **402**, 282 (2010)
33. D.T. Maltby et al., MNRAS **402**, 282 (2010)
34. B. Milvang-Jensen et al., A&A **482**, 419 (2008)
35. B. Moore, G. Lake, N. Katz, ApJ **495**, 139 (1998)
36. R. Pelló et al., A&A **508**, 1173 (2009)
37. B.M. Poggianti et al., ApJ **642**, 188 (2006)
38. B.M. Poggianti et al., ApJ **684**, 888 (2008)
39. B.M. Poggianti et al., ApJ **693**, 112 (2009)
40. B.M. Poggianti et al., The Messenger **136**, 54 (2009)
41. G. Rudnick et al., ApJ **700**, 1559 (2009)
42. G. Rudnick et al., The Messenger **112**, 19 (2003)
43. R.P. Saglia et al., A&A **524**, A6 (2010)
44. P. Sánchez-Blázquez et al., A&A **499**, 47 (2009)
45. L. Simard et al., A&A **508**, 1141 (2009)
46. B. Vulcani et al., MNRAS **412**, 246 (2011)
47. I.M. Whiley et al., MNRAS **387**, 1253 (2008)
48. S.D.M. White et al., A&A **444**, 365 (2005)
49. C. Wolf et al., MNRAS **393**, 1302 (2009)

The HAWK-I Cluster Survey

M. Huertas-Company, C. Lidman and The HCS collaboration

Abstract Distant galaxy clusters allow us to study the processes that drive galaxy evolution in the densest environments of the Universe. At these distances, near-IR observations are essential, as the rest frame optical is redshifted into the near-IR. However, with one or two exceptions, there is a distinct lack of high-quality imaging data at these wavelengths. To overcome this limitation, we have obtained a uniform set of deep near-IR images of ten of the most distant galaxy clusters currently known. Among the main goals of the project there are the detailed study of the evolution of the CMR, the scaling relations and merger rates of cluster galaxies between $z \sim 0.8$ and $z \sim 1.45$. We summarize here the current status of the survey and presents a first look at some of the data.

1 Survey Description

Our HAWK-I sample consists of a subsample of ten clusters that were targeted by the SCP (Supernova Cosmology Project) together with the addition of two other clusters (Table 1). The redshift range of the clusters is broad ($0.84 < z < 1.45$ covering 2 Gyr of cosmic time) and each cluster contains between 10 and 100 spectroscopically confirmed cluster members. All of the clusters, except RDCS J1252-2927, were targeted with HAWK-I. HAWK-I is ESOs wide-field near-IR imager on the VLT. Its wide field-of-view, high throughput and excellent image

M. Huertas-Company (✉)
Paris Observatory, 5 Place Jules Janssen, 92195 Meudon, France
and
University of Paris 7, 75205, Paris cedex 13, France
e-mail: marc.huertas@obspm.fr

C. Lidman
Australian Astronomical Observatory, 167 Vimiera Rd, Eastwood, NSW 2122, Australia
e-mail: clidman@aao.gouv.au

I. Ferreras and A. Pasquali (eds.) *Environment and the Formation of Galaxies: 30 years later*, Astrophysics and Space Science Proceedings,
DOI 10.1007/978-3-642-20285-8_31, © Springer-Verlag Berlin Heidelberg 2011

Table 1 Summary of clusters in the sample

Cluster selection	Redshift	Selection	Comment
RX J0152-1357	0.84	X-ray	Not part of the SCP sample
RCS 2319+0028	0.91	Optical	
XMM J1229+0151	0.98	X-ray	
RCS 0220.9-0333	1.03	Optical	
RCS 2345-3633	1.04	Optical	
XMM J0233-0436	1.22	X-ray	
RDCS J1252-2927	1.23	X-ray	Observed with ISAAC
SpARCS J0035-4312	1.34	Spitzer	Not part of the SCP sample
XMMU J2235.3-2557	1.39	X-ray	
XMM J2215-1738	1.45	X-ray	

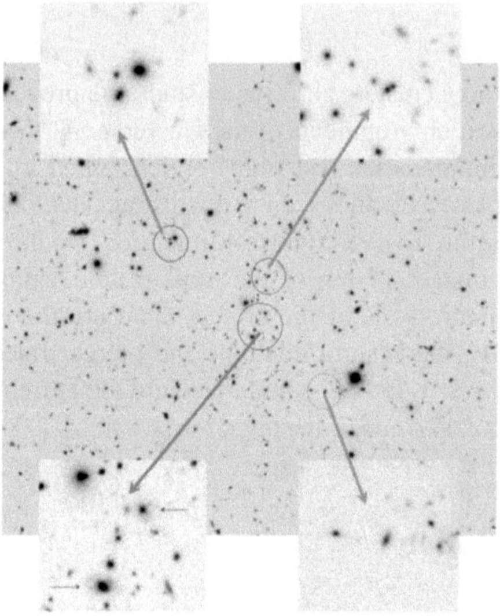

Fig. 1 SpARCS J0035-4312 was discovered in the Spitzer Adaptation of the Red Sequence Cluster Survey (SpARCS) [1, 3, 4]. Based on the cluster red-sequence technique of Gladders and Yee [2], SpARCS extends the redshift limit of this technique to higher redshifts by combining ground based optical and Spitzer [3.6] and [4.5] data. With a redshift of $z = 1.34$, SpARCS J0035-4312 is one of the most distant clusters currently known. Panels highlight close pairs (*upper panels*), evidence for major mergers (*lower left panel*) and a very dusty and IR luminous edge on disk galaxy (*lower right panel*)

quality combine to make it one of the most powerful groundbased near-IR images. Our program was completed earlier this year. More information can be found in http://hcs.obspm.fr. The data are now being processed. An initial inspection of the data shows that the image quality is excellent, with values as low as 0.25 arcsec in the fully processed images. The image quality is good enough to morphologically classify galaxies, and to look for evidence of major merging, as we will demonstrate with SpARCS J0035-4312, which is the only cluster in the sample lacking deep ACS data (Fig. 1).

References

1. R. Demarco et al., ApJ **711**, 1185 (2010)
2. M.D. Gladders, H.K.C. Yee, AJ **120**, 2148 (2000)
3. A. Muzzin et al., ApJ **698**, 1934 (2009)
4. G. Wilson et al., ApJ **698**, 1943 (2009)

3D Spectroscopy Unveils Massive Galaxy Formation Modes at High-z

F. Buitrago, C.J. Conselice, B. Epinat, A.G. Bedregal, I. Trujillo, R. Grützbauch and The GNS team

Abstract Massive (stellar mass $\geq 10^{11} M_\odot$) galaxies at high redshift ($z \geq 1.5$) remain mysterious objects. Their extremely small sizes (effective radii of $1-2$ kpc) make them as dense as globular clusters, whereas in the present day Universe similar mass systems are large with old and metal-rich stellar populations. In order to explore this development, we present near-IR IFU observations with SINFONI@VLT for ten massive galaxies at $z \sim 1.4$ solely selected by their high stellar mass which allows us to retrieve velocity dispersions, kinematic maps and dynamical masses. We join this with imaging from the GOODS NICMOS Survey (GNS), which was carried out by our group, and which is the largest sample of massive galaxies (80 objects) at high redshift ($1.7 < z < 3$) to date. With these data we show how their morphology changes, possibly as a result of minor merging events also seen in the kinematics.

1 On the Nature of Massive Galaxies

In the era of high redshift galaxy detection, massive galaxies are not well understood, partially due to their striking compactness ([1], see Fig. 1 left side) and also their number densities are not reproduced by current galaxy formation models [3].

F. Buitrago (✉) · C.J. Conselice · R. Grützbauch
School of Physics and Astronomy, University of Nottingham NG7 2RD, UK
e-mail: ppxfb@nottingham.ac.uk

B. Epinat
Laboratoire d'Astrophysique de Toulouse-Tarbes, 31400, Toulouse, France

A.G. Bedregal
Departamento de Astrofísica, Universidad Complutense de Madrid, 28040, Madrid, Spain

I. Trujillo
Instituto de Astrofísica de Canarias, 38205, Tenerife, Spain

I. Ferreras and A. Pasquali (eds.) *Environment and the Formation of Galaxies: 30 years later*, Astrophysics and Space Science Proceedings,
DOI 10.1007/978-3-642-20285-8_32,

Although the formation mechanism of massive galaxies is not perfectly established [8] it is clear they should become the cores of the most massive galaxies in the local Universe [5]. Major merging and AGN feedback cannot account by themselves for their mass and size growth, due to the noticeable star formation seen in these compact objects [7], also observed in the far infrared [2, 9]. As such, the most probable mechanism for this galaxy population to evolve is the continuous bombardment of minor objects belonging to their surroundings [6]. In order to constrain this scenario beyond pure photometry we present here integral field spectroscopy for a number of these massive galaxies.

2 3D Spectroscopy Massive Galaxies Sample

Apart from all the studies we performed in the GOODS NICMOS Survey ([4], see www.nottingham.ac.uk/astronomy/gns for a list of all the projects of the GNS team), our group was granted 20 h observation time utilizing SINFONI 3D spectrograph at ESO-VLT. In Fig. 1 right side, we show several of our kinematical maps to be soon published. Our aim is not only the discovery of merging in this sample, but also to constrain spectroscopically whether these objects are better represented by a spheroid or a disk galaxy population. We are currently developing models to fully characterize their rotation, but there are hints of an increase in the number of massive disks at high redshift, witnessing also several minor mergers.

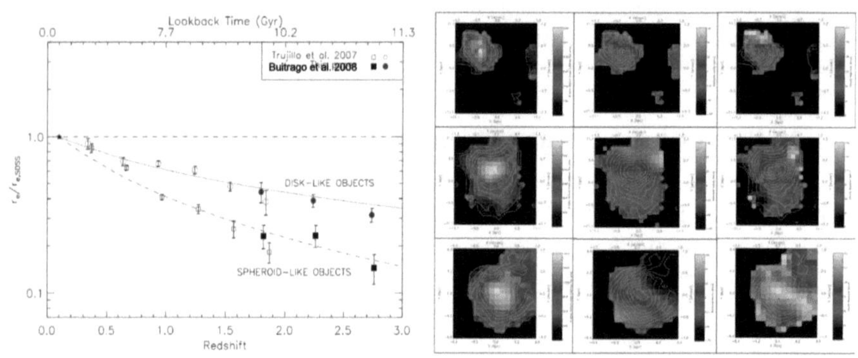

Fig. 1 *Left side:* The size evolution of massive galaxies ($M_* > 10^{11}\,M_\odot$) with redshift [1]. Plotted is the ratio of the median sizes of galaxies in our sample with respect to average sizes of nearby galaxies in the SDSS local comparison. The *error bars* indicate the uncertainty (1σ) at the median position. *Right side:* Here we present a sample of three galaxies (*rows*) from our SINFONI massive sample, which comes from the POWIR Survey detection and DEEP2 spectroscopy redshift determination. Columns from *left* to *right* are H_α flux, radial velocity and projected velocity dispersion maps. All the pixels shown are above $S/N = 3$

References

1. F. Buitrago et al., ApJ **687**, L61 (2008)
2. A. Cava et al., MNRAS **409**, L19 (2010)
3. C.J. Conselice et al., MNRAS **381**, 962 (2007)
4. C.J. Conselice et al., MNRAS **413**, 80 (2011)
5. P.F. Hopkins et al., MNRAS **398**, 898 (2009)
6. T. Naab et al., ApJ **699**, L178 (2009)
7. P.G. Pérez-González et al., ApJ **687**, 50 (2008)
8. E. Ricciardelli et al., MNRAS **406**, 230 (2010)
9. M. Viero et al., (2010). arXiv: 1008.4359

Detailed Stellar Population Analysis of Galaxy Clusters at Increasing Redshift: A Constrain for Their Evolution

A. Ferré-Mateu and P. Sánchez-Blázquez

Abstract This contribution describes a project of a detailed stellar population analysis for different clusters of galaxies with increasing redshift. The selected clusters are selected from the *GEMINI/HST Galaxy Cluster Project*. GMOS spectra and HST imaging are avaiable for all of them. The signal-to-noise ratio of the spectra allow us to analyse the stellar population of individual Galaxies, allowing us to study trends with morphological type or distance to the center of the cluster. Here we intend to show the different techniques we are planning to apply in this study to derive kinematics, ages, metallicities and star formation histories for the idividual galaxies in each clusters. Furthermore, an analysis of all this properties with the redshift is planned to constrain the evolution of the galaxies.

1 Data Sample and Methology

The *GEMINI/HST Galaxy Cluster Project* (PIs I. Jørgensen and R. Davies) consists in 15 X-ray selected galaxy clusters with redshifts between 0.2 and 0.9. The sample is limited to clusters with X-ray luminosities larger than $L_x = 2\times 1{,}044\ \mathrm{ergs}^{-1}$ and is intended to be representative of rich galaxy clusters. The galaxies for which spectra were obtained can be several times less luminous than the Milky Way. The spectra have a very high signal-to-noise (S/N) ratio, making it possible to analyze stellar populations in individual galaxies. We consider three redshift bins: low-z ($z < 4$), mid-z ($0.4 \leq z \leq 0.7$) and hihg-z ($z > 0.7$), which allows us to constrain the galaxy evolution by studying the stellar populations in early-type galaxies. At the

A. Ferré-Mateu (✉)
Instituto de Astrofísica de Canarias, 38205 La Laguna, Spain
e-mail: aferre@iac.es

P. Sánchez-Blázquez
Universidad Autónoma de Madrid, 28049 Madrid, Spain
e-mail: psanchezblazquez@googlemail.com

I. Ferreras and A. Pasquali (eds.) *Environment and the Formation of Galaxies: 30 years later*, Astrophysics and Space Science Proceedings,
DOI 10.1007/978-3-642-20285-8_33,

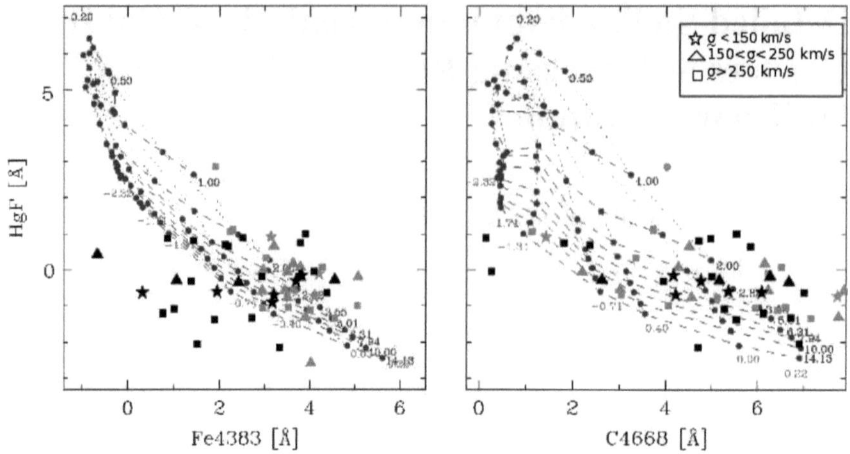

Fig. 1 The age-sensitive indicator $H\gamma_F$ is plotted against some metallicity indices. The SSP model grids of Vazdekis et al. [6] are plotted. Age increases from *top* to *bottom* and metallicity from *left* to *right*. *Black symbols* correspond to the mid-z cluster and *grey symbols* to the high-z one

moment three different clusters are being analyzed, each one corresponding to one redshift bin. The proposed clusters are reduced following the standard reduction process (bias substraction, flat-fielding, cosmic ray removal, wavelenght calibration, telluric lines removal and flux calibration. The velocity dispersions and radial velocities are derived by means of the Penalized Pixel-Fitting method `PPxF` [1], having previously fitted the emission lines with `GANDALF` [5]. After this, the line strenght indices are measured and compared with the new Vazdekis' stellar population models [6] and with the routine `RMODEL` (provided by N. Cardiel), ages and metallicities for each galaxy are derived (see Fig. 1, where a typical index-index grid is shown just for examplification, presenting the values of 49 galaxies from one cluster at z = 0.55 and another at 0.83). Finally, several techniques of full spectral fitting to derive the star formation history of each galaxy are applied. We are using `ULySS` [3], `STARLIGHT` [2] and `STECKMAP` [4]. Once this porcess is carried out for individual galaxies in each cluster, we will study the evolution of some elements with redshift, to prove formation and evolution theories.

2 Future Work

Although each cluster is studied independently of the others by different coauthors, there are many other topics that can be addressed with these galaxies:

- Study of the Fundamental Plane
- Study the environmental dependence
- Jeans modelling
- Study of the true nature of ellipticals vs S0

References

1. M. Capellari, E. Emsellem, PASP **116**, 138 (2004)
2. R. Cid-Fernandes et al., MNRAS **358**, 363 (2005)
3. M. Koleva et al., A&A **501**, 1269 (2009)
4. P. Ocvirk et al., MNRAS **356**, 74 (2006)
5. M. Sarzi et al., MNRAS **366**, 1151 (2006)
6. A. Vazdekis et al., MNRAS **404**, 1639 (2010)

The Role of Galaxy Stellar Mass in the Colour–Density Relation up to $z \sim 1$

O. Cucciati, A. Iovino, K. Kovač and zCOSMOS Collaboration

Abstract It is well known that galaxy properties correlate with the local environment in which galaxies reside. In contrast, it is still matter of debate why and when these environmental dependences originate, and whether only one "main" property depends on environment, thus driving all the others environmental dependences via the correlations among properties themselves. We use the first ~10,000 spectra of the zCOSMOS sample ($I \leq 22.5$) to study the role of galaxy stellar mass in the colour–density relation up to $z = 1$. We confirm that within a luminosity-limited sample ($M_B \leq -20.5 - z$) red galaxies reside mainly in high densities (δ) at least up to z = 1. This trend becomes weaker for increasing redshifts, and it is mirrored by the D_n4,000 Å break–density relation. We also find that up to $z \sim 1$ the fraction of galaxies with log(EW[OII]) ≥ 1.15 is higher for lower δ. Given the triple dependence among galaxy colours, stellar mass and δ, the colour–δ relation that we find can be due to the broad range of stellar masses embedded in the sample. We find that once mass is fixed the colour-δ relation is globally flat up to $z \sim 1$ for galaxies with $\log(M/M_{sun}) \geq 10.7$. On the contrary, even at fixed mass we observe that within $0.1 \leq z \leq 0.5$ the fraction of red galaxies with $\log(M/M_{sun}) \leq 10.7$ depends on δ. We suggest a scenario in which the colour depends primarily on stellar mass, but for an intermediate mass regime ($10.2 \leq \log(M/M_{sun}) \leq 10.7$) the local density

O. Cucciati (✉)
Laboratoire d'Astrophysique de Marseille, UMR 6110 CNRS-Université de Provence, 38 rue Frederic Joliot-Curie, 13388 Marseille Cedex 13, France
e-mail: olga.cucciati@oamp.fr

A. Iovino
INAF-OABrera, Via Brera 28, 20121 Milano, Italy
e-mail: angela.iovino@brera.inaf.it

K. Kovač
Institute of Astronomy, ETH Zürich, 8093, Zürich, Switzerland
e-mail: kovac@phys.ethz.ch

I. Ferreras and A. Pasquali (eds.) *Environment and the Formation of Galaxies: 30 years later*, Astrophysics and Space Science Proceedings,
DOI 10.1007/978-3-642-20285-8_34,

modulates this dependence. These lower mass galaxies formed more recently, in an epoch when evolved structures were already in place, and their longer SFH allowed environment-driven physical processes to operate during longer periods of time.

1 Data and Analysis

Data. zCOSMOS [12] is a large spectroscopic survey in the COSMOS field [17], conceived with the main goal of studying high-redshift galaxy environments from galaxy groups scales up to larger-scale structures. Here we make use of the first ~10,000 spectra of the zCOSMOS *Bright* sample ("10k-sample", $I \leq 22.5$). See references in [17] for details on multi-wavelength coverage of the COSMOS field.

Details about the zCOSMOS survey can be found in [12] and [13]. In summary, spectra have been collected with the VIMOS-VLT multi-spectrograph, using the Red MR grism. The 10k-sample has an average sampling rate of ~33%, a measurement redshift error of $\sim 100\,\mathrm{km\,s^{-1}}$, and it covers an area of $\sim 1.4\,\mathrm{deg^2}$. We refer the reader to [6] and references therein for details about the computation of absolute magnitudes, stellar masses, the equivalent width of the [OII]λ3727 doublet (*EW*[OII]) and the amplitude of the 4,000 Å break (D_n4,000 from now on).

The detailed description of the environment parametrisation within the zCOSMOS 10k-sample is given in [11]. Summarising, the environment surrounding a given galaxy at comoving position **r** has been characterised by the dimensionless density contrast $\delta(\mathbf{r}) = [\rho(\mathbf{r}) - \rho_m(\mathbf{r})]/\rho_m(\mathbf{r})$, where density is smoothed with a given filter. In this formula, $\rho_m(\mathbf{r})$ is the mean density at the redshift that corresponds to the comoving position **r**, while $\rho(\mathbf{r})$ is the local density around the given galaxy. As an optimal choice for this work, we use a density contrast estimator based on the 5th nearest neighbour (n.n.) method, computed in slices of $\pm 1{,}000\,\mathrm{km\,s^{-1}}$, with luminosity-limited tracers.

Analysis. As a first step, we exploited the broad redshift range covered by the zCOSMOS 10k-sample to investigate the evolution with cosmic time of environmental effects on galaxy properties. For this analysis, we used a luminosity-selected volume-limited sample ($M_B \leq -20.5 - z$). We found, confirming previous results, that the fraction of red ($U - B \geq 1$) galaxies depends on environment at least up to $z \sim 1$, with red galaxies residing preferentially in high density environments. This dependence becomes weaker for higher redshifts. We also found that the D_n4000-density relation mirrors the colour–density relation (as both red colour and high D_n4000 are indicators of old and passively evolving galaxies), and this is the first time that this is shown up to $z \sim 1$. In contrast, we found that the 'active' galaxies (those with larger *EW*[OII]) reside preferentially in low densities. As for colour and D_n4000, the *EW*[OII]-density relation holds up to $z \sim 1$, although with less significance, and weakens for increasing z. These results for colour and D_n4000 are shown in Fig. 1.

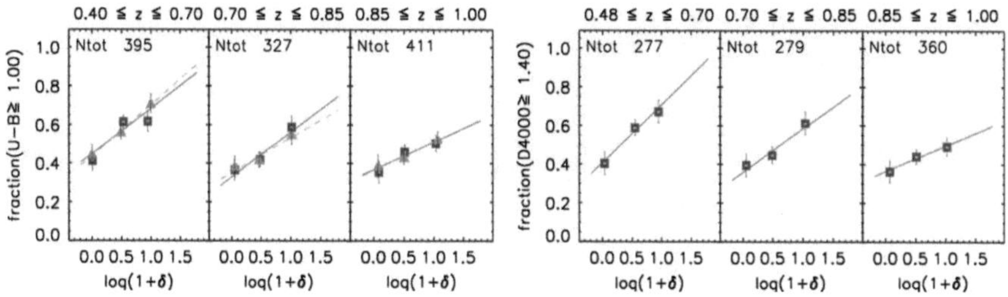

Fig. 1 *Left*. Fraction of red galaxies as a function of the density contrast. Different symbols are for different density measurement methods (see [6]). The *solid line* is the linear fit of *squares*, and the *dashed line* is for *triangles*. 'Ntot' is the total number of galaxies used in each panel. *Right*. Fraction of 'passive' ($D_n4000 \geq 1.40$.) galaxies as a function of the density contrast

Given the fact that stellar mass depends on local density (see for example [2]), and that colour depends on mass, it is difficult to interpret the meaning of the colour–density relation in a luminosity-selected sample. In fact, the wide spread in mass-to-light ratios embedded in this selection leave us with a broad range in stellar masses.

The only way to disentangle the three-fold colour-mass-density relation is using galaxy subsamples at fixed stellar mass, to avoid any dependence of mass on local density. We studied the colour–density relation in mass bins of $\Delta \log(M/M_{\odot}) = 0.2$, in three different redshift bins, attaining to the mass limits imposed by the flux limit of our survey. Our results are shown in Figs. 2 and 3. Note that for this analysis we use a different definition of 'red' galaxies: see the red line in the upper panel of Fig. 3). We found that, once the mass is fixed, the colour–density relation is globally flat, but with some exceptions. We observe that within $0.1 \leq z \leq 0.5$ the fraction of red galaxies depends on environment even when mass is fixed, at least for $\log(M/M_{\odot}) \leq 10.7$. This means that environment affects directly not only the stellar mass, but also other galaxy properties, at least for these given mass and redshift ranges.

2 Discussion

It is nowadays well assessed that in the local Universe, on average, redder and brighter galaxies live preferentially in regions where the local density is higher, while the opposite is true for bluer and fainter galaxies (for example, [1, 10]). Globally, this picture is still valid at higher redshift ($z \leq 1$). Although it is commonly observed that the pictures at $z \sim 0.1$ and at $z \sim 1$ are linked by the progressive weakening of the environmental effects for increasing redshift [4, 5], it is not still totally clear when this environmental dependence was established.

Unfortunately, the evolution of the environmental effects on galaxy properties up to $z \sim 1$ and above has always been studied in luminosity selected samples.

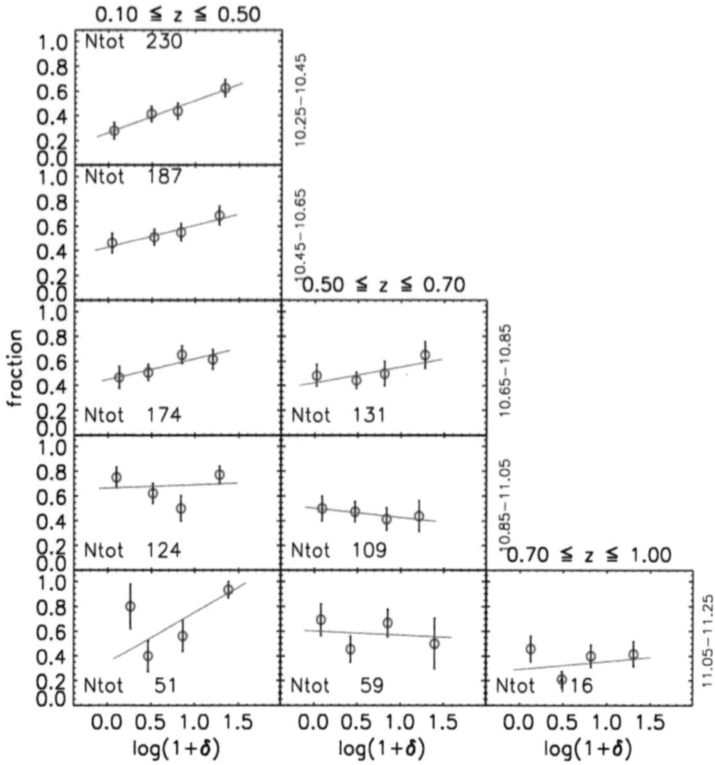

Fig. 2 Colour–density relation in three redshift bins (columns, redshift bins quoted on the *top*) and five mass bins (*rows*, mass bins quoted on the right, in logarithmic scale). *Circles* show the fraction of galaxies with $U - B$ colour redder than a threshold that depends on the stellar mass, as described in Fig. 3. *Solid lines* represent the linear fit of *circles*. In each panel, 'Ntot' is the total number of galaxies in that mass and redshift bin

We have shown above that this selection may bring to a misleading interpretation, given the underlying mass-density relation. For example, the faster decreasing with redshift of the red fraction in high densities than in low densities, found in both [4] and [5], can be interpreted as the proof that 'nurture' is at work, differently shaping galaxy properties in different environments. But when we take into account the mass-density relation, these findings become easily interpretable with a biased galaxy formation: more massive galaxies formed first in the highest density peaks (for example [15]), and being more massive they consumed their gas reservoir more quickly, causing the fast increase of the red fraction with cosmic time in high densities.

In the past years it has been shown that the galaxy stellar mass drives the star formation history of galaxies (for example [7]), that in turn is mainly responsible for the galaxy colour. Moreover, the stellar mass is related on the one side to the galaxy halo mass [14], and on the other side to the local environment as determined by

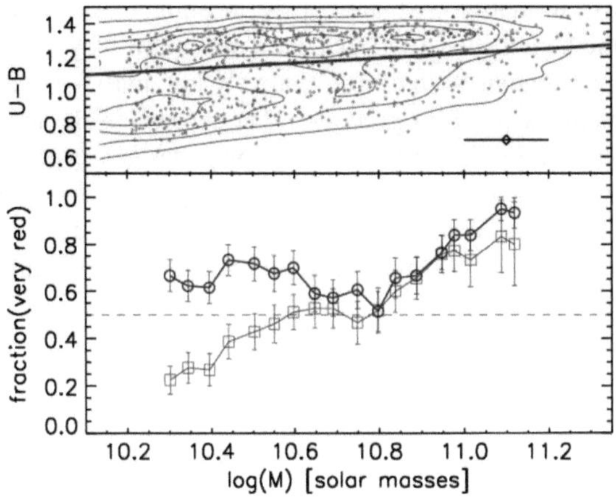

Fig. 3 *Top*: the colour-mass plane for galaxies in the range $0.1 \leq z \leq 0.5$, with the diagonal thick line being the threshold used to define 'red' galaxies, meant to be roughly parallel to the red sequence in the colour-mass plane and 0.3 magnitudes bluer than the red sequence itself. *Bottom*: the fraction of 'red' galaxies in the highest δ quartile (*circles*) and in the lowest δ quartile (*squares*) for not-independent bins of $\Delta \log(M) = 0.2$

nearby galaxies [2,10]. It follows that a non-biased analysis of environmental effects on galaxy colour (or on other properties) can be addressed only within complete mass-selected subsamples at fixed stellar mass.

In our analysis we find that, once the mass is fixed, the colour–density relation is globally flat, but in the redshift range $0.1 \leq z \leq 0.5$ there exists a sort of threshold mass ($\log(M/M_{\odot}) \sim 10.7$), below which the mix of the two broad galaxy populations (red and blue, early and late) does still depend on local density once stellar mass is fixed. Accordingly to these results, we asses that the colour depends primarily on mass, but for the low-mass regime the local environment modulates this dependence, playing a direct role in shaping galaxies properties.

A possible general picture to describe these findings is the following. Considering that the reddest and most massive galaxies have been formed earlier (at $z > 2$, having already an age of >1 Gyr at $z \sim 1$, see for example [18] and references therein), and that moreover their Star Formation Histories (SFH) are typically faster (≤3 Gyr, see for example [8]) than those of lower mass galaxies, their formation epoch and their evolution time-scale took place on the mean before an appreciable growth of structures (at $z \sim 1$ the number of structures with $\log(M/M_{\odot}) > 5 \times 10^{14}$ was ~100 times lower than today, in the concordance cosmology, see [3]). We do not see evident environmental effects on these galaxies because environment could not affect their evolution. On the contrary, lower mass galaxies ($\log(M/M_{\odot}) \leq 10.7$) not only have been formed more recently, but their SFH is also slower, and thus it could have been modified by physical processes typical of high density regions.

This picture is in agreement with the scenario where both 'nature' and 'nurture' affect galaxy properties (e.g. [5, 10]), to which we can add the dependence on mass. First, galaxy formation is 'biased' [15], meaning that more massive galaxies formed first in the highest density peaks, later on followed by lower mass galaxies in less dense environment. Second, for relatively low mass galaxies the evolution is affected by complex physical processes that depend on the local environment. According to this picture, the properties of a low-mass galaxy are thus affected by the environment in which the galaxy resides at any given epoch of its evolution, and not only at the time of its formation. We refer the reader also to [16] for a global picture, based on zCOSMOS data, on the role of mass and environment in influencing galaxy evolution.

References

1. M.R. Blanton et al., ApJ **629**, 143 (2005)
2. M. Bolzonella et al., A&A **524**, 76 (2010)
3. S. Borgani, Lecture Notes in Physics, Springer, **740**, 287 (2008)
4. M.C. Cooper et al., MNRAS **376**, 1445 (2007)
5. O. Cucciati et al., A&A **458**, 39 (2006)
6. O. Cucciati et al., A&A **524**, 2 (2010)
7. G. Gavazzi, M. Scodeggio, A&A **312**, 29 (1996)
8. G. Gavazzi et al., ApJ **576**, 135 (2002)
9. A. Iovino et al., A&A **509**, A40+ (2010)
10. G. Kauffmann et al., MNRAS **353**, 713 (2004)
11. K. Kovač et al., ApJ **708**, 505 (2010)
12. S.J. Lilly et al., ApJS **172**, 70 (2007)
13. S.J. Lilly et al., ApJS **184**, 218 (2009)
14. R. Mandelbaum et al., MNRAS **368**, 715 (2006)
15. C. Marinoni et al., A&A **442**, 801 (2005)
16. Y. Peng et al., ApJ **721**, 193 (2010)
17. N. Scoville et al., ApJS **172**, 1 (2007)
18. D. Vergani et al., A&A **487**, 89 (2008)

Selection of Luminous Galaxies at the Edge of the Universe

S. Pereira de Matos, J. Afonso, H. Messias, and C. Santos

Abstract The search and study of distant galaxies is crucial for the understanding of the very first stages in galactic evolution. One way to find such systems is by selecting sources strongly emitting at radio frequencies. Radio-loud galaxies can be promising distant candidates since they can be detected everywhere in the Universe with the current deep radio surveys. However, the best radio-galaxy high-z candidates may well be weakly or even not detected at other wavelengths. For this purpose, we address a sample of double-lobe candidate radio sources detected at 1.4 GHz by the Australian Telescope Compact Array (ATCA) in the Extended *Chandra* Deep Field South ($\alpha = 3^h32^m30^s, \delta = -27°48^m20^s$ (J2000)). This region has also been surveyed by *Spitzer Space Telescope* in the infrared (IR) and by the Advanced Camera for Surveys (on board of *Hubble Space Telescope*) in the optical, both observations reaching enormous depths. Here, the results from comparing these two wavebands with the ATCA radio map are presented, aiming for a more restrict sample of high-z candidates. This study will improve with the advent of future facilities such as the Squared Kilometer Array or the Atacama Large Millimetre Array, providing deeper and better resolution radio/mm surveys or follow-up observations.

1 Introduction

Powerful high-z radio galaxies (HzRGs) may be easily detected based on their morphology, revealing kpc-long emission jets originating from the vicinity of the central supermassive black hole, thus enabling the detection at very high redshifts. This project aims at selecting luminous HzRGs, from a morphological point of

S.P. de Matos (✉) · J. Afonso · H. Messias · C. Santos
Observatório Astronómico de Lisboa, Centro de Astronomia e Astrofísica da Universidade de Lisboa, Tapada da Ajuda 1349 – 019 Lisbon, Portugal
e-mail: smatos@oal.ul.pt

I. Ferreras and A. Pasquali (eds.) *Environment and the Formation of Galaxies: 30 years later*, Astrophysics and Space Science Proceedings,
DOI 10.1007/978-3-642-20285-8_35,

Table 1 Summary of the radio, optical and IR surveys

Survey	Area	Bands	Limiting sensitivity	Reference
ATCA	1.2 deg^2	1.4 GHz	15 μJy (rms)	[1]
GOODS[a]	300 $arcmin^2$	z_{850}; V_{606}	26.6; 27.8 (AB, 10 σ)	[2]
GEMS[b]	800 $arcmin^2$	z_{850}; V_{606}	27.1; 28.3 (AB, 5 σ)	[4]
SWIRE[c]	6 deg^2	3.6; 4.5; 5.8; 8; 24 μm	5; 9; 43; 40; 193 μJy (5σ)	[3]

[a] Great observatories origin deep survey
[b] Galaxy evolution from morphology and SEDs
[c] Spitzer wide-area infrared extragalactic

view, in highly sensitive surveys using data collected at different wavelengths (see Table 1).

2 Selection of the Candidates

The galaxies addressed in this study were selected based on their radio morphology, where those visually resembling two radio jets were initially considered. This accounts for 94 candidate sources.

3 Optical/IR Identifications

The galaxy producing the radio emission may then be temptatively detected by overlaying the radio contours on the optical and IR images. Since the optical images have a much better resolution and are so deep, random associations between the radio sources and the optical ones are much more likely than in the IR. Furthermore, detection of galaxies in the IR wavelengths is more successful than in the optical and can provide a first indication of a galaxy's redshift [1]. Nonetheless, several radio sources remain unidentified either in the optical or the IR surveys.

Among the initial 94 candidate sources singled out in the ATCA survey, 14 have an IR counterpart, of which only seven likely correspond to the galaxy emitting two jets at radio frequencies. Within the area covered either by GEMS or GOODS, 7 out of 40 sources have a candidate optical counterpart. A more conservative analysis reduces the IR and optical candidate identifications to 4 and one counterparts, respectively, being these the rightful candidates for high-z radio sources.

References

1. J. Afonso et al., AJ **131**,1216 (2006)
2. M. Giavalisco et al., ApJ **600**, L93 (2004)
3. R.P. Norris et al., AJ **132**, 2409 (2006)
4. H.-W. Rix et al., (2004). arXiv:astro-ph/0401427v1

From Fields to a Super-Cluster: The Role of the Environment at $z = 0.84$ with HiZELS

D. Sobral, P. Best, I.J. Smail, and HiZELS team

Abstract At $z = 0$, clusters are primarily populated by red, elliptical and massive galaxies, while blue, spiral and lower-mass galaxies are common in low-density environments. Understanding how and when these differences were established is of absolute importance for our understanding of galaxy formation and evolution, but results at high-z remain contradictory. By taking advantage of the widest and deepest Hα narrow-band survey at $z = 0.84$ over the COSMOS and UKIDSS UDS fields, probing a wide range of densities (from poor fields to rich groups and clusters, including a confirmed super-cluster with a striking filamentary structure), we show that the fraction of star-forming galaxies falls continuously from ~40% in fields to approaching 0% in rich groups/clusters. We also find that the median SFR increases with environmental density, at least up to group densities – but only for low and medium mass galaxies, and thus such enhancement is mass-dependent at $z \sim 1$. The environment also plays a role in setting the faint-end slope (α) of the Hα luminosity function. Our findings provide a sharper view on galaxy formation and evolution and reconcile previously contradictory results at $z \sim 1$: stellar mass is the primary predictor of star formation activity, but the environment also plays a major role.

1 Introduction

Star formation activity is strongly dependent on environment: clusters of galaxies are primarily populated by passively-evolving galaxies, while star-forming galaxies are common in low-density environments [5]. It is well-established [1] that the typical star formation rates of galaxies – and the star-forming fraction – decrease

D. Sobral (✉) · P. Best · I.J. Smail
Institute for Astronomy, ROE, Blackford Hill, Edinburgh, UK
e-mail: drss@roe.ac.uk

I. Ferreras and A. Pasquali (eds.) *Environment and the Formation of Galaxies: 30 years later*, Astrophysics and Space Science Proceedings,
DOI 10.1007/978-3-642-20285-8_36, © Springer-Verlag Berlin Heidelberg 2011

with local galaxy density (often projected local density, Σ) both in the local Universe and at moderate redshift [10]. Active star-forming galaxies in the local Universe are also found to have lower masses than passive galaxies and, indeed, the most massive galaxies are mostly non-star-forming, an observational result often known as mass-downsizing [4]. While massive galaxies are predominantly found in high density environments, it has been shown that the mass-downsizing trend is not simply a consequence of the environmental dependence, nor vice-versa [13].

When did the environment and mass dependences of star-forming galaxies start to be observable, and how did they affect the evolution of galaxies and clusters? By $z \sim 1$, some authors [3, 6] have claimed to have found a flattening or even a definitive reverse of the relation between star formation activity and local galaxy density (SFR-Σ relation). However, other studies [7, 12, 14] argue that even at $z \sim 1$ both star formation rate and the star-forming fraction decline with increasing local density. Part of the discrepancies may be due to mass dependences already in place at high redshift. In order to identify and distinguish the separate roles of mass and environment on star formation at high redshift, one really requires clean, robust and large samples of star-forming galaxies residing in a wide range of environments and with a wide range of masses, together with samples of the underlying population found at the same redshift. Narrow-band Hα surveys are one of the most effective ways to gather representative samples of star-forming galaxies at different epochs, and the scientific potential of these is now being widely explored, following the development of wide-field cameras in the near-infrared. In particular, HiZELS, the Hi-Redshift(z) Emission Line Survey [9, 15], is playing a world-leading role by obtaining very large samples of Hα emitters (and other emission lines; see [16]), at $z = 0.84$, $z = 1.47$ and $z = 2.23$ (see [2] for an overview) over various square degree fields with a wealth of high-quality multi-wavelength data.

2 Samples and Properties

This study uses the large sample of Hα emitters at $z = 0.84$ from HiZELS presented in [15], modified as detailed in [18]. Briefly, the sample was derived from a narrow-band J filter ($\Delta\lambda = 0.015\,\mu$m) survey using WFCAM/UKIRT and reaching a star formation rate (SFR) limit in Hα of $3\,M_{\odot}\,yr^{-1}$ over 1.3 deg^2 in the UKIDSS UDS and the COSMOS fields. Photometric redshifts were used to select a sample which is $> 95\%$ reliable and complete (based on zCOSMOS – see [15]); this has been further modified to (1) take into account highly improved photometric redshifts, (2) reject potential AGN identified in [8] and (3) include EW< 50 Å Hα emitters (c.f. [18]). The final sample contains a total of 770 star-forming Hα emitters.

The improved high quality photometric redshifts at $z \sim 0.8$ available in COSMOS and UDS are used to emulate the narrow-band filter selection and to

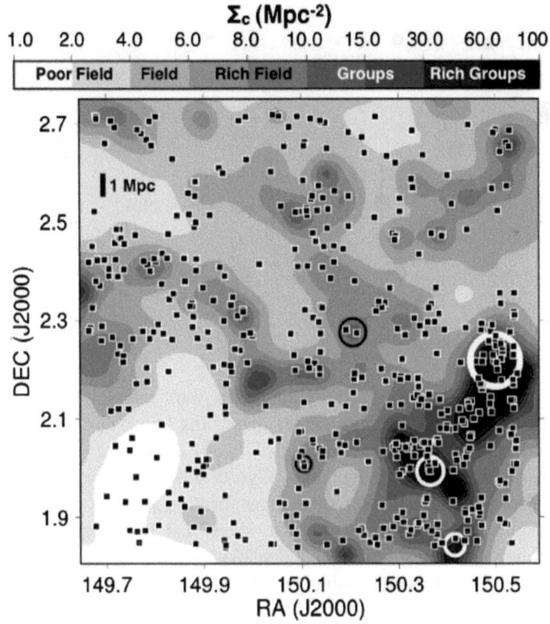

Fig. 1 The on-sky distribution of star-forming Hα emitters at $z = 0.84$ in the COSMOS field (*filled squares*), compared with the local projected density field (grayscale; using a 10th nearest neighbor analysis). *Circles* mark the positions of extended X-ray emission from confirmed groups and clusters within the narrow-band redshift range, scaled to reflect the measured X-ray luminosity (in a log scale). Lower X-ray luminosity groups are identified with *black circles*, while richer X-ray groups are plotted in white for good contrast. Note the very rich structure in the COSMOS field (including a rich cluster), providing a unique opportunity to probe the densest environments

select the underlying population. Spectroscopic redshifts provide completeness and contamination estimates of various photometric-redshift selected samples, allowing for appropriate corrections to be made. The underlying sample contains 6,344 sources.

Stellar masses are determined using a detailed SED fitting (with a wide range of parameters and fixing $z = 0.84$), and environmental densities are based on a 10th nearest neighbor analysis, corrected for completeness and contamination by using spectroscopic redshifts (c.f. [18], including detailed studies showing how the results are robust against errors and systematics). Local environmental densities are classified using the real-space correlation length, using the analysis presented in [17] into *fields* ($\Sigma_c < 10\,\mathrm{Mpc}^{-2}$), *groups* ($10 < \Sigma_c < 30\,\mathrm{Mpc}^{-2}$) and *rich groups/clusters* ($\Sigma_c > 30\,\mathrm{Mpc}^{-2}$). A very wide range of environments is probed, confirmed by the detection of extended X-ray emission in the highest density regions (see Fig. 1).

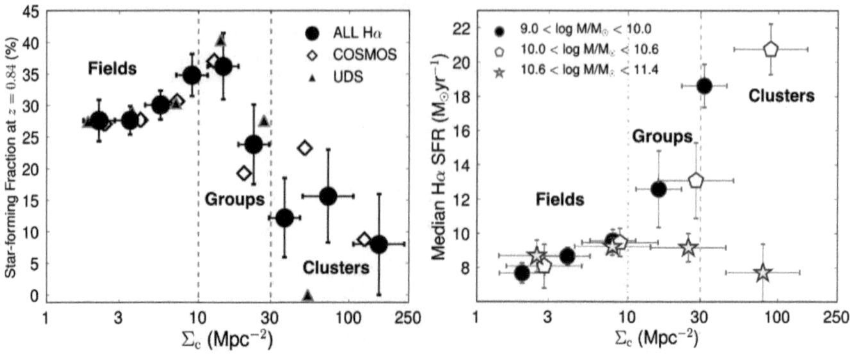

Fig. 2 The fraction of the Hα star-forming galaxies (*left panel*) and the median Hα star formation rate of Hα emitters in different stellar mass bins (*right panel*), as a function of local projected galaxy surface density (SFRs > $3\,M_\odot\,yr^{-1}$). At the lowest densities, the star-forming fraction is relatively constant and only increases slightly with Σ_c. However, for higher densities, $\Sigma_c > 10\,\mathrm{Mpc}^{-2}$, there is a steep decline of the star-forming fraction down to the highest (rich groups/cluster) densities probed. The results also show that the typical (median) SFR of Hα emitters increases continuously from the lowest densities to group densities, but this is only found for low/medium masses: massive star-forming galaxies have SFRs which are mostly unaffected by their environment

3 Environmental Dependences at Z ∼ 1

We find that the fraction of galaxies forming stars (above the HiZELS limit) is relatively flat with increasing local density within the field regime, but it falls sharply with density once group densities are reached (Fig. 2), resulting in a fall from the field to rich groups/clusters, as seen in the nearby Universe and consistent with [7, 12]. The median star formation rates of star-forming galaxies increases with environmental density for both field and group environments (Fig. 2), in good agreement with [3, 6], but the trend is stopped for the highest densities, where it is appears to fall (c.f. [18]), also agreeing with studies probing such very high densities [14]. Furthermore, we find that the environment changes the shape of the Hα luminosity function: the faint-end slope (α) is found to vary with environmental density, being very steep ($\alpha \sim -2$) for poor fields, and very shallow for the highest density regions (groups and clusters, $\alpha \sim -1$), as shown in Fig. 3.

4 Mass-Environment View at Z ∼ 1

We find that mass-downsizing is fully in place at $z \sim 1$, with the fraction of star-forming galaxies declining steeply with stellar mass (c.f. [18]). Since stellar mass and environment are correlated, to which extent could the environmental trends be driven by stronger, more fundamental mass trends? In order to address this

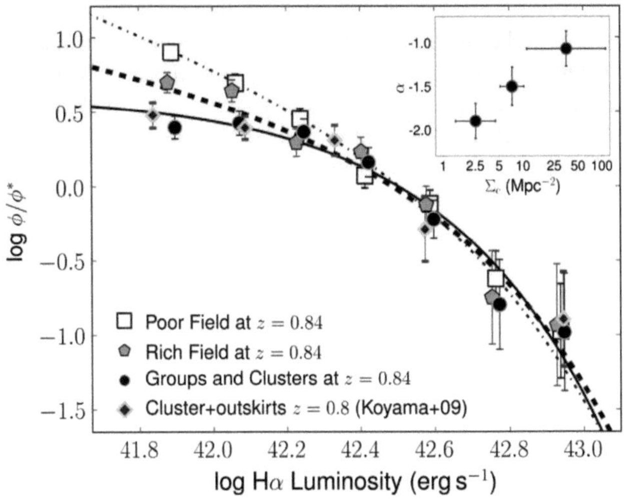

Fig. 3 The normalised Hα luminosity function shows a significant difference in the faint-end slope, α, as a function of local surface density. As shown in the inset, α is very steep ($\alpha \approx -1.9$) for poor field, shallower for medium densities (rich field), and very shallow for groups and clusters, with $\alpha \approx -1.1$. Recent results from [11] fully agree with the Hα luminosity function derived for similar local projected densities

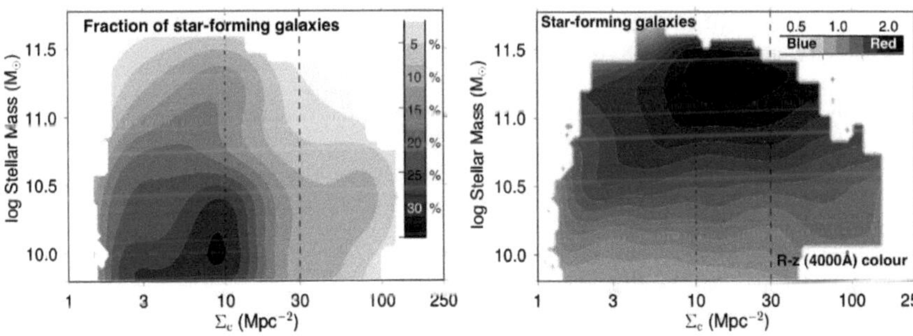

Fig. 4 *Left*: The fraction of star-forming galaxies (mergers excluded) as a function of both mass and environment, revealing that both are important at $z \sim 1$, just like in the local Universe. *Right*: The distribution of the median R-z colour (roughly probing the 4,000 Åbreak colour at $z = 0.84$) within the mass-density 2D space for the Hα star-forming galaxies; this reveals that stellar mass is the main colour predictor, as the environment only correlates weakly with colour

question, we have taken advantage of the large samples to investigate dependences on both mass and environment, simultaneously. Interestingly, when potentially merger-driven star-formation (which dominates at high densities) is neglected, we find that the fraction of star-forming galaxies declines independently with both environment and stellar mass (fixing the other), clearly showing no qualitative evolution from the local Universe (Fig. 4). However, we also find that the $\sim$4,000 Å break colour of star-forming galaxies depends almost uniquely, and very strongly, on stellar mass (Fig. 4), and not on environment. Interestingly, we also find that the positive SFR-Σ correlation is driven by low and median stellar mass galaxies: the median star formation rates of the most massive galaxies are mostly unaffected by the environment, and sSFRs of such massive galaxies actually decline with Σ; this fully explains apparently contradictory results in the literature which used different selections and probed different environments.

Overall, we find that stellar mass is the primary predictor of star formation activity at $z \sim 1$, but the environment, while initially enhancing the star formation activity of (lower-mass) star-forming galaxies, is ultimately responsible for suppressing star-formation activity in all galaxies above surface densities of groups and clusters.

References

1. P.N. Best MNRAS **351**, 70 (2004)
2. P.N. Best et al., UKIRT@30 Proceedings (2010) arXiv:1003.5183
3. M.C. Cooper et al., MNRAS **383**, 1058 (2008)
4. L.L. Cowie et al., AJ **112**, 839 (1996)
5. A. Dressler, ApJ **236**, 351 (1980)
6. D. Elbaz et al., MNRAS **468**, 33 (2007)
7. R.A. Finn et al., ApJ **630**, 206 (2005)
8. T. Garn et al., MNRAS **402**, 2017 (2010)
9. J. Geach et al., MNRAS **388**, 1473
10. T. Kodama et al., MNRAS *354*, 1103 (2004)
11. Y. Koyama et al., MNRAS **403**, 1611 (2010)
12. S.G. Patel et al., ApJL **705**, L67 (2009)
13. Y. Peng et al., ApJ **721**, 193 (2010)
14. B.M. Poggianti et al., ApJ **684**, 888 (2008)
15. D. Sobral et al., MNRAS **398**, 75 (2009)
16. D. Sobral et al., MNRAS **398**, L68 (2009)
17. D. Sobral et al., MNRAS **404**, 1551 (2010)
18. D. Sobral et al., MNRAS **411**, 675 (2011)

Witnessing a Link Between Starburst and AGN Activities at $2 < z < 4$?

H. Messias, J. Afonso, B. Mobasher, A. Hopkins, and D. Farrah

Abstract One of the main subject of study in modern astronomy is the field of galaxy evolution, how mass has been assembled since the earliest age of the Universe. Therefore, it is of fundamental necessity the ability to pinpoint key evolution stages in the galaxy population. Although much work has been done in isolated phases of galaxy evolution (starburst, passive, AGN, ...), the ultimate goal is actually the study of galaxies transiting in between main phases of galaxy evolution, allowing us to know the sequence of events and reasons behind. Here, a candidate sample for such scenario is presented. Although only scarce references are found in the literature regarding this population, it is already accepted that both old and hot young stellar populations co-exist in these systems. In addition, optical/nIR spectroscopy and X-ray data also point to the presence of obscured AGN activity.

H. Messias (✉) · J. Afonso
Centro de Astronomia e Astrofísica da Universidade de Lisboa, Observatório Astronómico de Lisboa, Tapada da Ajuda, 1349-018 Lisbon, Portugal
e-mail: hmessias@oal.ul.pt; jafonso@oal.ul.pt

B. Mobasher
University of California, University Av, Riverside, CA 92508, USA
e-mail: mobasher@ucr.edu

A. Hopkins
Anglo-Australian Observatory, Epping, NSW 1710, Australia
e-mail: ahopkins@aao.gov.au

D. Farrah
University of Sussex, East Sussex BN1 9RH, UK
e-mail: D.Farrah@sussex.ac.uk

I. Ferreras and A. Pasquali (eds.) *Environment and the Formation of Galaxies: 30 years later*, Astrophysics and Space Science Proceedings,
DOI 10.1007/978-3-642-20285-8_37,

1 Sample

Referred to as pure distant red galaxies (pDRGs) in [7], these galaxies found in the *Chandra* Deep Survey South (CDFS) are selected by a simple colour–colour criterion: $J - K_s > 1.3$ and $i - K_s < 2.5$ (AB). The majority of the sample (91%) falls at $2 < z < 4$, an epoch of peak activity in the Universe. The objects are selected from the FIREWORKS catalogue [12] to a depth of $K_{s,TOT} = 24.3$ AB, resulting in a 111 sources population (101 at $2 < z < 4$). As an example of the peculiarity of pDRGs, the BzK criterion [2] fails to properly classify many of these sources labeling them as $z < 1.4$ sources, when spectroscopy confirms such galaxies at $z > 2$.

2 SFRs, AGN Content, and Morphology

Previous and current work using radio data [7] assignes a star formation rate (SFR) upper limit of $100\,M_\odot\,yr^{-1}$. The rest-frame UV excess (UVxs) of pDRG enables a detection at optical wavebands and implies SFRs of just a few to as high as $\sim$100 $M_\odot\,yr^{-1}$. Although dust seems not to be abundant in pDRGs, it may still be affecting UV light. Considering dust in spectral energy distribution fitting procedures, SFRs as high as 500–800 $M_\odot\,yr^{-1}$ can be found in a couple of examples. If this is the case, some pDRGs can produce $z \sim 1$ massive systems of 3–4$\times 10^{11} M_\odot$, considering observed masses [6, 8] and an exponentially decaying SF history ($\tau = 0.3$–1 Gyr).

Both spectroscopy and X-rays show evidences of obscured AGN activity, meaning a SF nature of the UVxs seen in pDRGs. Among these objects, one can find the type-2 QSO announced by [9] as the farthest at the time ($z = 3.7$), an object showing P-cygni profiled high ionization lines (NV, CIV, SiIV; [1]), and a FeLoBAL (e.g., [3]) candidate. Fitting semi-empirical models [10, 11] to the overall rest-frame photometric data, implies a 20–30% AGN contribution to the IR. The best fit templates are based in a known local ULIRG merger system, IRAS 22491-1808.

Testing for a merger induced scenario the morphology extraction code developed by [4, 5] was used and the cut $M_{20} > -1.1$ was adopted as a strong indicator for clumpy/irregular profiles, hence merging system candidates. Separating the galaxy sample at $2 < z < 4$ into non-DRGs, DRGs not pDRGs, and pDRGs (hence complementary samples), the M_{20} cut indicates, respectively, clumpy systems fractions of 34, 36, and 46%. A strong evidence for the perturbed nature of pDRGs.

A final conclusion can not yet be established. More spectroscopic deep coverage of the sample is most definitely necessary, and the 4Ms *Chandra* observations on CDFS are now being analysed. Still, the evidences presented here strongly point to the co-existence of mixed stellar populations with obscured nuclear activity.

References

1. S. Cristiani et al., A&A **359**, 489 (2000)
2. E. Daddi et al., ApJ **617**, 746 (2004)
3. M. Greg et al., ApJ **573**, 85 (2002)
4. J. Lotz, J. Primack, P. Madau, AJ **128**, 163 (2004)
5. J. Lotz et al., ApJ **636**, 592 (2006)
6. D. Marchesini et al., ApJ **701**, 1765 (2009)
7. H. Messias et al., ApJ **719**, 790 (2010)
8. N.M. Förster Schreiber et al., ApJ **706**, 1364 (2009)
9. C. Norman et al., ApJ **571**, 218 (2002)
10. M. Polletta et al., ApJ **663**, 81 (2007)
11. M. Salvato et al., ApJ **690**, 1250 (2009)
12. S. Wuyts et al., ApJ **682**, 985 (2008)

Galaxy Properties in Different Environments at $z > 1.5$ in the GOODS-NICMOS Survey

R. Grützbauch, R.W. Chuter, C.J. Conselice, A.E. Bauer, A.F.L. Bluck, F. Buitrago, and A. Mortlock

Abstract We present a study of the relationship between galaxy colour, stellar mass, and local galaxy density in a deep near-infrared imaging survey up to a redshift of $z \sim 3$ using the GOODS NICMOS Survey (GNS). The GNS is a very deep, near-infrared Hubble Space Telescope survey imaging a total of 45 arcmin^2 in the GOODS fields, reaching a stellar mass completeness limit of $M_* = 10^{9.5}\ M_\odot$ at $z = 3$. Using this data we measure galaxy local densities based on galaxy counts within a fixed aperture, as well as the distance to the third, fifth and seventh nearest neighbour. We find a strong correlation between colour and stellar mass at all redshifts up to $z \sim 3$. We do not find a strong correlation between colour and local density, however, the highest overdensities might be populated by a higher fraction of blue galaxies than average or underdense areas, indicating a possible reversal of the colour-density relation at high redshift. Our data suggests that the possible higher blue fraction at extreme overdensities might be due to a lack of *massive* red galaxies at the highest local densities.

1 Introduction

Despite a wealth of recent studies, disentangling the influence of stellar mass and environment on galaxy formation and evolution remains a highly debated topic. The presence of a strong correlation between galaxy colour and stellar mass is supported by numerous observational studies in the local Universe (see e.g. [11, 12]) and up to intermediate redshifts of $z \sim 1$ (see e.g. [3, 9]). However, galaxy colours not only depend on mass, but also on a galaxy's environment. The preference of red galaxies for denser local environments, first noticed by [14], and confirmed by [7], is now

R. Grützbauch (✉) · R.W. Chuter · C.J. Conselice · A.E. Bauer · A.F.L. Bluck · F. Buitrago · A. Mortlock
University of Nottingham, Nottingham, UK
e-mail: ruth.grutzbauch@nottingham.ac.uk

I. Ferreras and A. Pasquali (eds.) *Environment and the Formation of Galaxies: 30 years later*, Astrophysics and Space Science Proceedings,
DOI 10.1007/978-3-642-20285-8_38, © Springer-Verlag Berlin Heidelberg 2011

well-studied in the local Universe (e.g. [13,17]). In the early Universe, however, the evidence is controversial. While some studies find that the colour-density relation at $z \sim 1$ is mainly due to a bias in stellar mass selection and only persists for low-mass galaxies (e.g. [9, 10, 16]), others argue that a strong colour-density relation is already in place at $z \sim 1.4$ (e.g. [5,8]) even at fixed stellar mass [6]. In this study we investigate the relationship between galaxy colours, stellar mass and local densities in the critical redshift range at $z > 1.5$, probing the period in which most of galaxy formation takes place.

2 Data and Analysis

We use data from the GOODS NICMOS Survey (GNS), a large Hubble Space Telescope survey reaching a stellar mass completeness limit of $M_* \sim 10^{9.5}\ M_\odot$ at $z \sim 3$ and covering a total area of about 45 arcmin^2 [4]. The limiting magnitude reached at 5σ is $H_{AB} = 26.8$, which is more than 2 mag deeper than ground based near-infrared imaging [15]. To obtain photometric redshifts, stellar masses and rest-frame colours, the NICMOS H-band sources were matched to a catalogue of optical sources in the GOODS ACS fields, yielding a *BVizH* photometric catalogue.

The photometric redshifts are measured with the HYPERZ code [1]. For the high redshift sample ($z > 1.5$) we obtain a photometric redshift uncertainty of $\sigma_{\Delta z/(1+z)} = 0.10$, while galaxies at $z < 1.5$ show a slightly lower, but still comparable scatter of $\sigma_{\Delta z/(1+z)} = 0.08$. The photometric redshift uncertainties are used in a set of 100 Monte Carlo simulations to estimate the uncertainties in local density. For this purpose we randomize the photometric redshift input according to the photo-z error, accounting for scattering in and out of the redshift range, as well as dealing separately with catastrophic outliers. The randomized photo-zs are then used to recompute the local densities, giving typical average uncertainties of ~0.2 dex. We plot the results of the simulations in Figs. 1 and 2 discussed in Sect. 3.2.

The stellar masses and $(U - B)$ rest-frame colours are measured by fitting a large set of synthetic spectra constructed from the stellar population models of [2], assuming a Salpeter initial mass function. The average uncertainty of our stellar mass estimate is ~0.3 dex.

We compute local densities using the fixed aperture method as well as the Nth nearest neighbour approach. The absolute densities are corrected for edge effects and normalized by the median value in a redshift bin of $\Delta z = 0.25$, yielding a relative local density $\log\ (1 + \delta_N)$. Comparing the different density estimators we conclude that the third nearest neighbour density has the largest dynamical range and only slightly higher uncertainties and is therefore best suited to trace the extremes in the local density distribution. We show the results for $\log\ (1 + \delta_3)$ in the following.

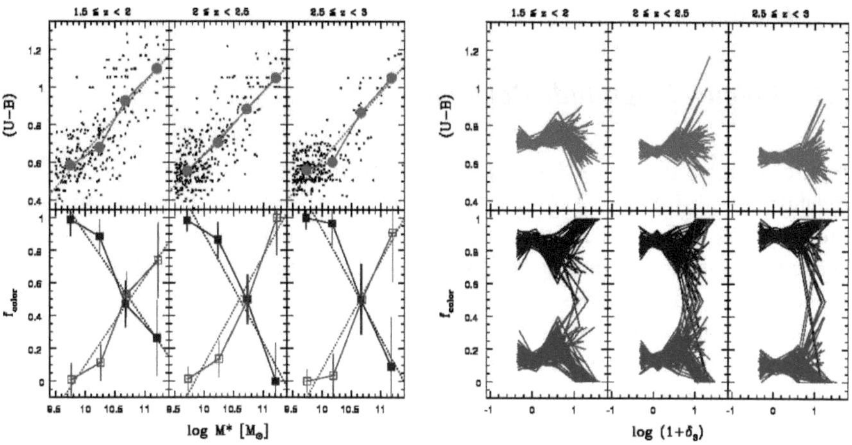

Fig. 1 $(U - B)$ colour and fraction of red and blue galaxies as a function of stellar mass (*left*) and local density log $(1 + \delta_3)$ (*right*). Each line corresponds to the results of one Monte Carlo Simulation, averaged in bins of stellar mass (*left*) and local density (*right*)

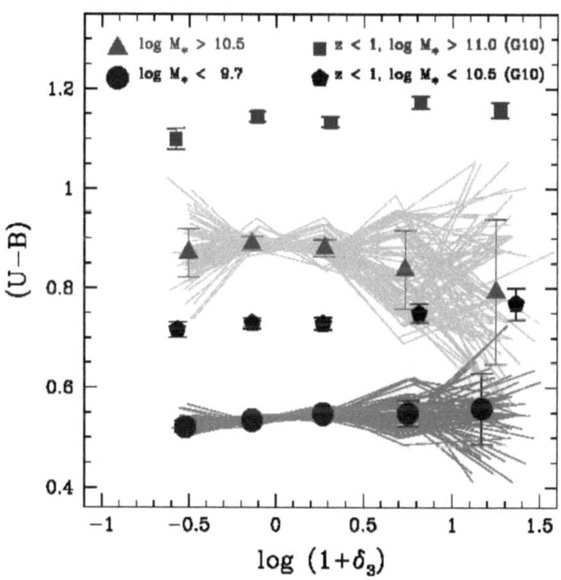

Fig. 2 Colour density relation in low (*black*) and high (*grey*) stellar mass quartiles over the whole redshift range $1.5 < z < 3$. The mean and RMS of all Monte Carlo runs in bins of local density are overplotted as grey triangles (high quartile) and black circles (low quartile). The average data points from the sample of [9] at intermediate redshift $(0.4 < z < 1)$ are plotted as big grey squares (high quartile) and grey pentagons (low quartile), respectively

3 Results

3.1 Colour–Magnitude Relation

Figure 3 shows the $(U - B)$ - M_B colour–magnitude relation in three different redshift bins: $1.5 < z < 2$, $2 < z < 2.5$ and $2.5 < z < 3$. In the three top panels the sample is split in 4 stellar mass bins with a bin size of 0.5 dex; the symbols are shaped and sized according to the galaxy's stellar mass: larger symbols represent higher stellar mass galaxies. In the three lower panels the sample is split in bins of local density with a bin size of 0.6 dex. Here, larger symbols correspond to galaxies located in higher overdensities. The red solid lines indicate the expected location of the red sequence observed at $z \sim 1$, and evolved passively back in time according to the models of [18]. The long-dashed line is the border to the blue cloud, defined as being 0.25 magnitudes bluer than the red sequence. The different slices in stellar mass are clearly separated in colour–magnitude space. The separation is perpendicular to the red sequence, and the distance from the expected red sequence increases with decreasing stellar mass. The different slices in local density on the other hand are not well separated in colour–magnitude space. The relation between colour and local density is much weaker than the relation between colour and stellar mass.

3.2 Colour–Density Relation and the Role of Stellar Mass

The left panel of Fig. 1 shows rest-frame $(U - B)$ colour (top) and the fractions of blue and red galaxies (bottom) as a function of stellar mass in three redshift bins. The correlation between $(U - B)$ colour and stellar mass log M_* is highly significant at all redshifts. The cross-over mass, i.e. the stellar mass at which there is an equal number of red and blue galaxies, stays roughly constant at log $M_* \sim 10.8$. This is the same cross-over mass found at lower redshifts of $0.4 < z < 1$ [9].

The right panel of Fig. 1 shows the relation between $(U - B)$ colour (top row) and the fraction of blue and red galaxies (bottom row) as a function of relative overdensity log $(1 + \delta_3)$. In this figure we plot the results of the Monte Carlo simulations rather than the original datapoints. Each line shows the average of one Monte Carlo run. We do not find a significant correlation between colour and overdensity, however, there is a possible trend for a higher fraction of blue galaxies ($\sim$100%) at the highest overdensities (log $(1 + \delta_3) > 0.8$) at all redshifts, where the blue fraction at intermediate and low densities is around 80–90%. Note that there is still a 10% probability that this trend is caused by chance.

Figure 2 shows the correlation between colour and local density for galaxies in the low and high stellar mass quartile. A clear colour offset between low- and high-mass galaxies of $\Delta(U - B) \sim 0.2 - 0.3$ mag is present at all densities. Interestingly, there is a trend that mean colours of galaxies in the high-mass quartile are bluer at

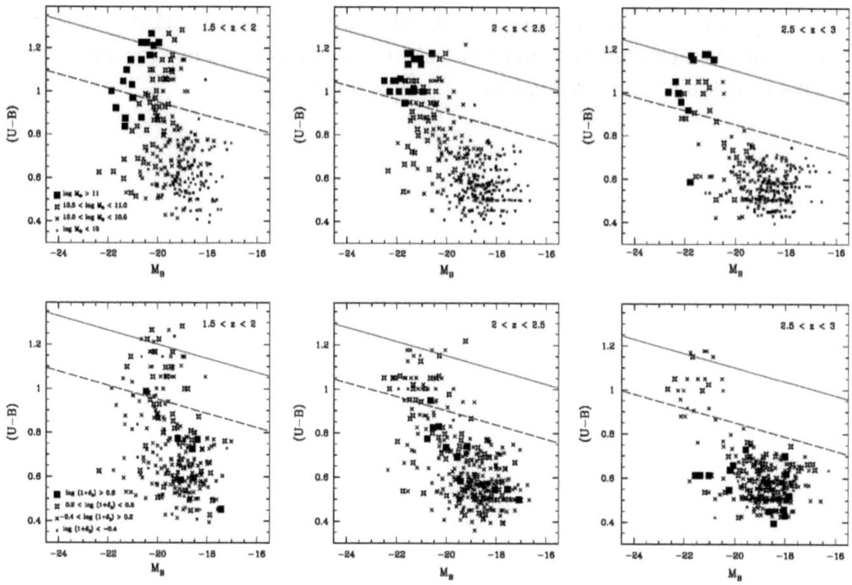

Fig. 3 Colour–magnitude relation in three redshift ranges, divided in bins of stellar mass (*top panels*) and local density (*bottom panels*). The larger symbols correspond to higher stellar mass in the *top panels*, and to higher overdensity in the bottom panels. The expected location of the red sequence at the respective redshift is shown as a *solid line*, while the limit to the *blue cloud* below which a galaxy is considered blue is plotted as a *long-dashed line*

higher overdensities ($\log\ (1 + \delta_3) > 0.5$) than at low and average local densities. For galaxies in the low-mass quartile we do not see a strong correlation between colour and local density.

4 Conclusions

In this study we investigate the influence of stellar mass and local density on galaxy rest-frame colour at a redshift of $1.5 < z < 3$. We find the following results:

- Galaxy colour depends strongly on galaxy stellar mass at all redshifts up to $z \sim 3$. After accounting for passive evolution in colour, the colour-stellar mass relation does not evolve with z below that redshift.
- Massive red galaxies exist up to $z \sim 3$, at the expected location of the red sequence in the colour–magnitude diagram.
- We do not find a correlation between colour and local density, however, galaxies in extremely high over-densities ($>5\times$ over-dense) are bluer than galaxies in average and most under-dense environments.

- More massive galaxies ($\log M_* > 10.5 M_\odot$) at high relative over-densities tend to be on average bluer than massive galaxies at average and low local densities. This is due to a lack of red galaxies at high over-densities.

References

1. Bolzonella, M., Miralles J.-M., & Pelló, R., 2000, A&A, 363, 476
2. Bruzual, G., & Charlot, S., 2003, MNRAS, 344, 1000
3. Bundy, K. et al., 2006, ApJ, 651, 120
4. Conselice, C.J. et al., 2010, preprint (arXiv:1010.1164)
5. Cooper, M.C. et al., 2006, MNRAS, 370, 198
6. Cooper, M.C. et al., 2010, MNRAS, 402, 1942
7. Dressler, A., 1980, ApJ, 236, 351
8. Gerke, B.F. et al., 2007, MNRAS, 376, 1425
9. Grützbauch, R. et al., 2010, MNRAS in press, arXiv:1009.3189
10. Iovino, A. et al., 2010, A&A, 509, 40
11. Kauffmann, G. et al., 2003, MNRAS, 341, 33
12. Kauffmann, G. et al., 2003, MNRAS, 341, 54
13. Kauffmann, G. et al., 2004, MNRAS, 353, 713
14. Oemler, A.Jr., 1974, ApJ, 194, 10
15. Retzlaff, J. et al., 2010, A&A, 511, 50
16. Tasca, L.A.M. et al., 2009, A&A, 503, 379
17. van der Wel, A., 2008, ApJ, 675, 13
18. van Dokkum, P.G., & Franx, M., 2001, ApJ, 553, 90

Simulations of Shell Galaxies with GADGET-2: Multi-Generation Shell Systems

K. Bartošková, B. Jungwiert, I. Ebrová, L. Jílková, and M. Křížek

Abstract As the missing complement to existing studies of shell galaxies, we carried out a set of self-consistent N-body simulations of a minor merger forming a stellar shell system within a giant elliptical galaxy. We discuss the effect of a phenomenon possibly associated with the galaxy merger simulations – a presence of multiple generations of shells.

1 Two-Generation Shell Structure

Galaxies with stellar shells are thought to be by-products of galaxy mergers [5]. Most, though not all, previous models, e.g. [2–5], relied on test-particle simulations. No systematic explorations of galactic models with these shell structures as merger debris via fully self-consistent N-body simulations, naturally involving the dynamical friction and the progressive decay of the accreted galaxy, were conducted. To bridge this gap, we decided to carry out a set of self-consistent simulations of a minor merger between a giant elliptical galaxy (gE with the mass of dark matter

K. Bartošková (✉)
Faculty of Science, Masaryk University, Brno, Czech Republic
and
Astronomical Institute, Academy of Sciences of the Czech Republic
e-mail: katka.bartoskova@gmail.com

B. Jungwiert · I. Ebrová · M. Křížek
Astronomical Institute, Academy of Sciences of the Czech Republic
and
Faculty of Mathematics and Physics, Charles University in Prague

L. Jílková
ESO Santiago, Chile
and
Faculty of Science, Masaryk University, Brno, Czech Republic

I. Ferreras and A. Pasquali (eds.) *Environment and the Formation of Galaxies: 30 years later*, Astrophysics and Space Science Proceedings,
DOI 10.1007/978-3-642-20285-8_39,

halo $8\times10^{12}\,M_\odot$ and stellar component $2\times10^{11}\,M_\odot$) and a satellite dwarf elliptical galaxy ($2\times10^{10}\,M_\odot$ in total), using the GADGET-2 code [6].

In order to study differences in the resulting shell system formed in differently centrally concentrated mass distributions, we prepared simulations with the gE galaxy in two versions: a two-component Plummer and a two-component Hernquist model, with the same effective radius. The dwarf galaxy is then released on a radial orbit with initial velocity $\sim$100 km s^{-1} and distance of 200 kpc from the giant galaxy.

In the first simulation, the core of the satellite passes through, returns and makes a second passage across the cent er of the primary galaxy ($\sim$1 Gyr after the first passage). This event leads to creating the second generation of shells. To our knowledge, this process has never been simulated in any previous study of the shell galaxies, although predictions in this sense were made, e.g. [1]. In the first approximation, we can look at this as a new collision between the returning core part of the satellite and the gE galaxy. Within the same generation, the shells of the debris system are moving with decreasing velocity. As the subsequent passage is not present in the latter simulation (with a two-component Hernquist model for the gE galaxy), the subsequent shells created after $\sim$1 Gyr move with different velocities compared to those belonging to the next generation in the former simulation, see Fig. 1.

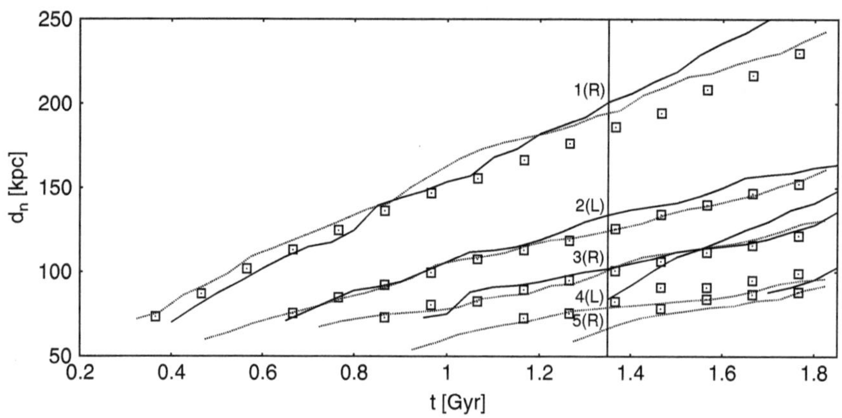

Fig. 1 The time evolution of the edges of three similar shell systems: results from the self-consistent simulation with gE galaxy modeled as the Plummer model (*solid lines*); from the test-particle simulation with the same initial conditions, but without the second passage of the satellite (*boxes*); and from the self-consistent simulation with gE galaxy represented by the Hernquist model (*dotted lines*). While the stars, gradually released from the potential of the satellite, oscillate in the potential of the gE galaxy, they are forming shells alternately interleaved on both sides of the gE galaxy along the merger axis – the first shell on the *right* side (R) is later followed by the second shell on the *left* (L), etc. The velocity of the shell-edge expansion – slope of $d_n(t)$ dependency – is given by the gravitational potential and by the shell ordinal number within a given generation. Therefore the first shells from each generation move with the same velocity. Such a shell galaxy arisen in the first self-consistent simulation, if observed in the particular time (e.g. 1.7 Gyr after the first passage), may appear to cause "the problem of a missing shell", since the third (originally fourth) outermost shell is then detected on the same side (L) as the second one

Acknowledgements This project is supported by grants No. 205/08/H005 (Czech Science Foundation), the project SVV 261301 (Charles University in Prague), MUNI/A/0968/2009 (Masaryk University in Brno), research plan AV0Z10030501 (Academy of Sciences of the Czech Republic) and LC06014 (Center for Theoretical Astrophysics, Czech Ministry of Education).

References

1. C. Dupraz, F. Combes, A&A **185**, 1 (1987)
2. I. Ebrová et al., ASPC **423**, 236 (2010)
3. L. Hernquist, P.J. Quinn, ApJ **331**, 682 (1988)
4. L. Jílková et al., ASPC **423**, 243 (2010)
5. P.J. Quinn, ApJ **279**, 596 (1984)
6. V. Springel, MNRAS **364**, 1105 (2005)

Comparing Various Approaches to Simulating the Formation of Shell Galaxies

I. Ebrová, K. Bartošková, B. Jungwiert, L. Jílková, and M. Křížek

Abstract The model of a radial minor merger proposed by [Quinn, ApJ 279, 596 (1984)], which successfully reproduces the observed regular shell systems in shell galaxies, is ideal for a test-particle simulation. We compare such a simulation with a self-consistent one. They agree very well in positions of the first generation of shells but potentially important effects – dynamical friction and gradual decay of the dwarf galaxy – are not present in the test-particle model, therefore we look for a proper way to include them.

1 Tides and Dynamical Friction in Test Particle Simulations

We model the luminous and dark matter components of a host giant elliptical (gE) and a dwarf elliptical galaxy by analytical potentials. In the simplest model, we assume the dwarf galaxy, filled with millions of test particles, to be ripped apart instantly when it comes close to the center of the gE galaxy. Its stars begin to oscillate in the potential of the host galaxy and produce shells at their turning points.

I. Ebrová (✉) · B. Jungwiert · M. Křížek
Astronomical Institute, Academy of Sciences of the Czech Republic
and
Faculty of Mathematics and Physics, Charles University in Prague
e-mail: ivana@ig.cas.cz

K. Bartošková
Astronomical Institute, Academy of Sciences of the Czech Republic
and
Faculty of Science, Masaryk University, Brno, Czech Republic

L. Jílková
ESO Santiago, Chile
and
Faculty of Science, Masaryk University, Brno, Czech Republic

I. Ferreras and A. Pasquali (eds.) *Environment and the Formation of Galaxies: 30 years later*, Astrophysics and Space Science Proceedings,
DOI 10.1007/978-3-642-20285-8_40, © Springer-Verlag Berlin Heidelberg 2011

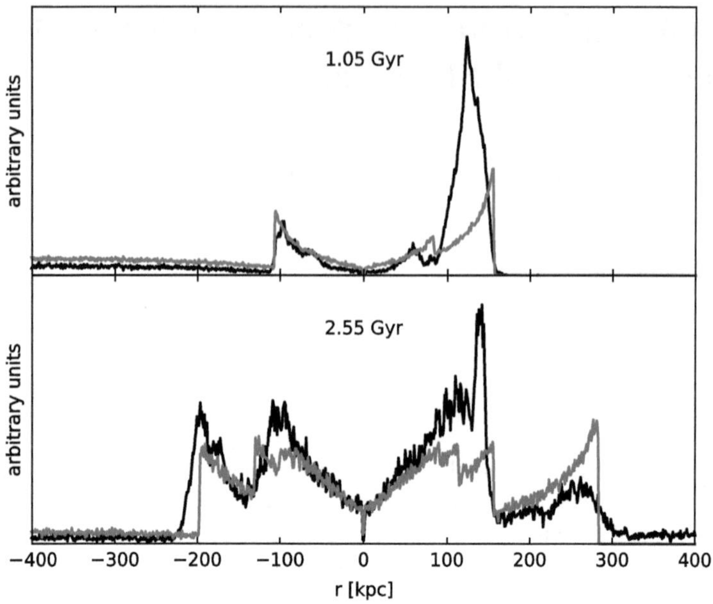

Fig. 1 The graph shows the comparison of histograms of radial distances of shells' particles (centered on the host galaxy) in the self-consistent (*black*) and test-particle (*grey*) simulations at two different time steps. Time equal zero corresponds to the passage of the dwarf galaxy through the center of the host galaxy. Both simulations have the same initial conditions. Notice that the brightness of the outermost shell (at 280 kpc at 2.55 Gyr) is supressed in the self-consistent simulation. This effect has been successfully simulated in the improved test-particle simulations; see [4]

Such a setup allows us to use large numbers of particles and so to gain sufficient contrast to detect all the shells in the simulation, also to investigate the kinematic footprint in spectral lines (see [2] and [5]) and explore a large parameter space. This would be very time consuming for large sets of self-consistent simulations.

Surprisingly, the agreement with a self-consistent simulation (for more details see [1]) turns out to be very good especially in the positions of shells (see Fig. 1 and [1]). But the simple model does not involve effects like dynamical friction and gradual decay of the dwarf galaxy, so that it cannot simulate phenomena seen in self-consistent simulations: the next generation of shells (see [1]) and the decrease of the shell brightness. We thus look for an intermediate way, where we can still have the large contrast available through the use of test particles, yet include some of the more complicated effects to make it more realistic.

In this improved test-particle simulation we use our version of enhanced Chandrasekhar formula with variable Coulomb logarithm to include dynamical friction, and we also introduced a gradual decline of the mass parameter of the dwarf galaxy potential to better imitate the evolution of tides. For details, see [3] and [4].

Acknowledgements We acknowledge the support by the grant No. 205/08/H005 (Czech Science Foundation), the project SVV 261301 (Charles University in Prague), EAS grant covering registration fee at JENAM 2010, the grant MUNI/A/0968/2009 (Masaryk University in Brno), the grant LC06014 'Center for Theoretical Astrophysics' (Czech Ministry of Education) and research plan AV0Z10030501 (Academy of Sciences of the Czech Republic).

References

1. K. Bartošková et al., this volume (2011) arXiv:1103.2562
2. I. Ebrová et al., this volume (2011) arXiv:1103.2563
3. I. Ebrová et al., (2011, in preparation)
4. I. Ebrová et al., ASPC **423** 236 (2010)
5. L. Jílková et al., ASPC **423**, 243 (2010)
6. P.J. Quinn, ApJ **279**, 596 (1984)

Modelling the Evolution of Galaxies as a Function of Environment

G. De Lucia

Abstract In this review, I provide an overview of theoretical aspects related to the evolution of galaxies as a function of environment. I discuss the main physical processes at play, their characteristic time-scales and environmental dependency, and comment on their treatment in the framework of hierarchical galaxy formation models. I briefly summarize recent results and the main open issues.

1 A Premise on 'Environment'

Historically, both theoretical and observational studies trying to assess the role of environment on galaxy evolution have been focused on galaxy clusters. One important reason for this is given by the practical advantage of having many galaxies in relatively small regions of the sky, and all approximately at the same distance. As laboratories to study galaxy evolution, however, clusters represent *rare* and *biased* systems: they originate from the highest peaks of the primordial density field, and evolutionary processes in these systems are expected to proceed at a somewhat accelerated pace with respect to regions of the Universe with 'average' density. Only about ten per cent of the cosmic galaxy population resides in clusters in the local Universe, and this fraction decreases with increasing redshift.

The recent completion of large spectroscopic and photometric surveys at different cosmic epochs has given new impetus to observational studies trying to assess the role of environment in galaxy evolution. Unfortunately, many different definitions of *environment* are used in the literature (e.g. some estimate of the halo mass, the number of neighbours counted in some volume, etc), depending on the available observational data, as well as on their quality. This unfortunate but

G. De Lucia (✉)
INAF – Astronomical Observatory of Trieste, via G. B. Tiepolo 11, 34143 Trieste, Italy
e-mail: delucia@oats.inaf.it

I. Ferreras and A. Pasquali (eds.) *Environment and the Formation of Galaxies: 30 years later*, Astrophysics and Space Science Proceedings,
DOI 10.1007/978-3-642-20285-8_41, © Springer-Verlag Berlin Heidelberg 2011

inevitable situation makes results from different surveys and/or at different cosmic epochs difficult to compare, and prevents them from putting really strong constraints on galaxy formation models.

In addition, it should be noted that in order to establish that physical processes related to a particular environment do play a role, it would be necessary to compare the evolution of *the same* galaxies in different environments. And this is certainly a difficult task from the observational viewpoint. Finally, one should take into account the hierarchical evolution of cosmic structures: dark matter collapses into haloes in a bottom up fashion, with small systems forming first and subsequently merging to form more massive structures. In this framework, galaxies might experience different 'environments' during their lifetimes. So, for example, galaxies residing in a cluster today might have suffered some degree of 'pre-processing' in lower mass systems like galaxy groups.

2 Methods

Hierarchical galaxy formation models find their seeds in the pioneering work by [33]. Galaxies are believed to originate from the condensation of gas at the centre of dark matter haloes: during the collapse of cosmic structure, gas is shock heated to very high temperatures and relaxes to a distribution that exactly parallels that of dark matter. Gas then cools, primarily via thermal Bremsstrahlung, and conservation of angular momentum leads to the formation of a rotationally supported disc. Mergers and instabilities form bulges, that can eventually grow a new disc, provided the system is fed by an appreciable cooling flow.

Different techniques are used to link the observed properties of luminous galaxies to those of the dark matter haloes in which they reside:

- In *semi-analytic models* of galaxy formation, the evolution of the baryonic components of galaxies is followed using simple yet physically and/or observationally motivated prescriptions to model complex physical processes like star formation, feedback, etc. Modern semi-analytic models take advantage of high-resolution N-body simulations to specify the location and evolution of dark matter haloes, which are assumed to represent the birth-places of luminous galaxies. This method can access large dynamic ranges in mass and spatial resolution, and allows a fast exploration of the parameter space, and an efficient investigation of different specific physical assumptions.
- *Direct hydrodynamical simulations* provide an explicit description of gas dynamics. This method is computationally more expensive then analytic models, so that large cosmological simulations are still somewhat limited by relatively low mass and spatial resolution. In addition, and perhaps most importantly, complex physical processes such as star formation, feedback, etc. still need to be modelled as *sub-grid physics*, either because the resolution of the simulation becomes

inadequate or because (and this is almost always true) we do not have a 'complete theory' for the particular physical process under consideration.
- *Halo Occupation Distribution (HOD) models* are based on a statistical characterization of the link between dark matter haloes and galaxies, bypassing any explicit modelling of the physical processes driving galaxy formation and evolution.

In the following, I will focus primarily on modelling of environmental processes through the first two methods mentioned above.

3 Physical Processes

Theoretically, there are a number of physical processes that can affect the evolution of galaxies in high density environments. In the following, I provide an overview of these processes, and describe how they are included in hierarchical galaxy formation models pointing out the main limitations and problems.

Galaxy mergers: Galaxy mergers and more generally strong galaxy-galaxy interactions, are commonly viewed as a rarity in massive clusters because of the large velocity dispersions of the systems. Mergers are, however, efficient in the infalling group environment, and may represent an important 'preprocessing' step in the evolution of cluster galaxies. Numerical simulations (see e.g. [1,6,19,28] and references therein) have shown that close interactions can lead to a strong internal dynamical response driving the formation of spiral arms and, in some cases, of strong bar modes. Sufficiently close encounters can completely destroy the disc, leaving a kinematically hot remnant with photometric and structural properties that resemble those of elliptical galaxies.

Mergers are intrinsically included in hierarchical galaxy formation models, and represent an important channel for the formation of bulges (see also Wilman, this volume). It should be noted that the number of 'important' mergers increases with increasing stellar mass, but is not as large as commonly thought [8]. In semi-analytic models, galaxy mergers are included adopting some variants of the classical Chandrasekhar dynamical friction formula. The left panel of Fig. 1 (from [9]) compares the merger times (in units of dynamical times) adopted in three different semi-analytic models with results from numerical simulations by [3]. The figure shows that, over the range of mass-ratios that provide merger times shorter than the Hubble time ($M_{\rm sat}/M_{\rm main} > 0.1$), there are large differences between different models. As discussed in [9], this has important consequences for the assembly history of massive galaxies, in particular of the brightest cluster galaxies (BCGs).

Controlled hydrodynamical simulations of merging disk galaxies are used to 'tune' the efficiency of merger driven starbursts in these models [6]. One important point, often neglected, is that these simulations are usually set-up using structural properties of nearby disk galaxies, which might not necessarily provide a good representation of systems merging at high redshift.

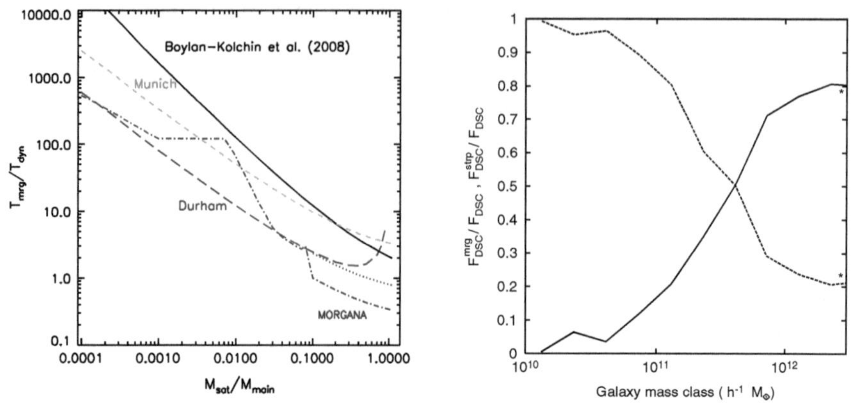

Fig. 1 *Left:* Merger times (in units of dynamical times) as a function of the baryonic mass ratio. *Dashed*, *long-dashed* and *dot-dashed lines* correspond to the standard assumptions adopted by the Munich, Durham and MORGANA models respectively. The *thick solid line* corresponds to the fitting formula provided by [3], for circular orbits. *Right:* The fraction of diffuse stars arising from mergers associated with the assembly history of the BCG (*solid line*), and with tidal stripping of stars from galaxies orbiting in the cluster potential (*dashed line*)

Recent theoretical work ([21] – their Fig. 8 is reproduced in the right panel of Fig. 1) suggests that merges, and more specifically those associated with the family tree of the BCG, are responsible for the largest fraction of the 'diffuse stellar component' that is observed in galaxy clusters. Semi-analytic models have only recently turned their attention to this component, and often model its formation only through tidal stripping [14, 27].

Gas stripping: Galaxies travelling through a dense intra-cluster medium suffer a strong ram-pressure stripping that can sweep cold gas out of the stellar disc [13]. Depending on the binding energy of the gas in the galaxy, the intra-cluster medium will either blow through the galaxy removing some of the diffuse interstellar medium, or will be forced to flow around the galaxy [5, 22]. Ram-pressure stripping is expected to be more important at the centre of massive systems because of the large relative velocities and higher densities of the intra-cluster medium. By considering the distribution and 'history of ram-pressure' experienced by galaxies in clusters, [4] estimated that strong episodes of ram-pressure are indeed predominant in the inner core of the clusters. They also showed, however, that virtually all cluster galaxies suffered weaker episodes of ram-pressure, suggesting that this physical process might have a significant role in shaping the observed properties of the entire cluster galaxy population. In addition, [4] found that ram-pressure fluctuates strongly so that episodes of strong ram-pressure alternate to episodes of weaker ram pressure, possibly allowing the gas reservoir to be replenished and intermittent episodes of star formation to occur.

While both numerical simulations and analytic studies show that ram-pressure stripping affects significantly the amount of gas in cluster galaxies, this physical process is usually not included in semi-analytic models of galaxy formation, with the exception of a couple of studies [15, 23]. These conclude that the inclusion of this physical process causes only mild variations in galaxy colours and star formation rates (I will explain why this is the case below – see 'strangulation' section). It should be noted that these studies include ram-pressure stripping using the original analytic formulation proposed by [13]. Recent numerical work [25] has shown that this formulation often provides incorrect mass-loss rates. In addition, the simple models used so far do not consider the possibility that ram-pressure stripping could temporarily enhance star formation [4, 29].

Strangulation: Current theories of galaxy formation assume that, when a galaxy is accreted onto a larger structure, the gas supply can no longer be replenished by cooling that is suppressed by the removal of the hot gas halo associated with the infalling galaxy [16]. This process is usually referred to as 'strangulation' (or 'starvation' or 'suffocation'). It is common to read in discussions related to these physical mechanisms, that strangulation is expected to affect the star formation of cluster galaxies on relatively long timescales, and therefore to cause a *slow* decline of the star formation activity. In semi-analytic models, however, this process is usually associated to a strong supernovae feedback, and is assumed to be instantaneous. As a consequence, galaxies that fall onto a larger system consume their cold gas rapidly, moving onto the red-sequence on very short time-scales (see e.g. [30]). This is also why ram-pressure stripping is not found to have a significant influence in these models: galaxies turn red and dead so quickly that ram-pressure does not have enough time to affect their colours or star formation rates.

Numerical simulations have recently pointed out that the stripping of the hot halo associated with infalling galaxies should not happen instantaneously. The left panel from Fig. 2 is reproduced from [18] and shows the evolution of the dark matter (solid line) and gas component (dashed line) associated with a galaxy placed on a realistic orbit through a cluster (see original paper for details). This particular simulation shows that the galaxy can retain ∼30% of its initial hot gas even ∼10 Gyr after

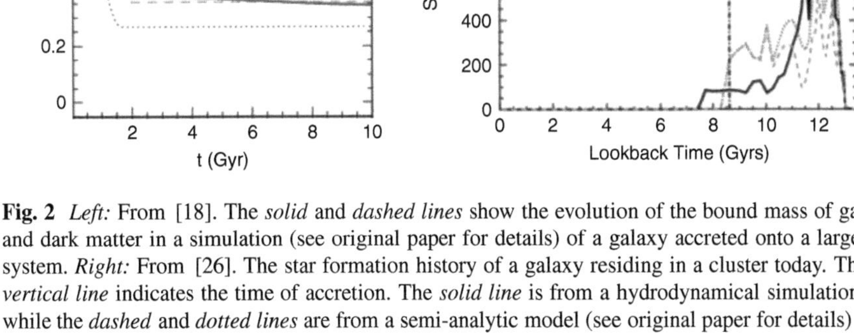

Fig. 2 *Left:* From [18]. The *solid* and *dashed lines* show the evolution of the bound mass of gas and dark matter in a simulation (see original paper for details) of a galaxy accreted onto a larger system. *Right:* From [26]. The star formation history of a galaxy residing in a cluster today. The *vertical line* indicates the time of accretion. The *solid line* is from a hydrodynamical simulations while the *dashed* and *dotted lines* are from a semi-analytic model (see original paper for details)

accretion. The right panel is reproduced from [27] and compares the star formation history of a cluster satellite galaxy from hydrodynamical simulations and semi-analytic models. [27] find that some cooling occurs on satellites and that this can last up to ~1 Gyr after accretion, but this seems to be important only for the most massive satellites. These simulations provide important (albeit still inconclusive) inputs for a revised treatment of this physical process in the framework of semi-analytic models of galaxy formation (see e.g [11, 31]).

Harassment: Galaxy harassment is a process that is not usually included in semi-analytic models of galaxy formation. The process has been discussed in early work on the dynamical evolution of cluster galaxies [24], and has been explored in detail using numerical simulations [10, 20]. These have shown that repeated fast encounters, coupled with the effects of the global tidal field of the cluster, can drive a strong response in cluster galaxies. The efficiency of the process is, however, largely limited to low-luminosity hosts, due to their slowly rising rotation curves and their low-density cores. Therefore, it is believed that harassment might have an important role in the formation of dwarf ellipticals or in the destruction of low-surface brightness galaxies in clusters [17], but it is less able to explain the evolution of luminous cluster galaxies.

AGN heating: Since the milestone paper by [32], it has been realized that some physical process is needed to suppress cooling flows at the centre of relatively massive haloes. Early semi-analytic models introduced ad-hoc prescriptions to suppress cooling flows in haloes above a critical mass. Modern models have included more accurate and physically motivated prescriptions, and have confirmed that AGN heating is indeed important to reproduce the exponential cut-off at the bright end of the galaxy luminosity function, and the old stellar populations observed for massive galaxies [8]. Figure 3 shows, for example, the predicted K-band luminosity function from two recently published semi-analytic models [2, 7] with and without AGN feedback (solid and dashed lines, respectively). The

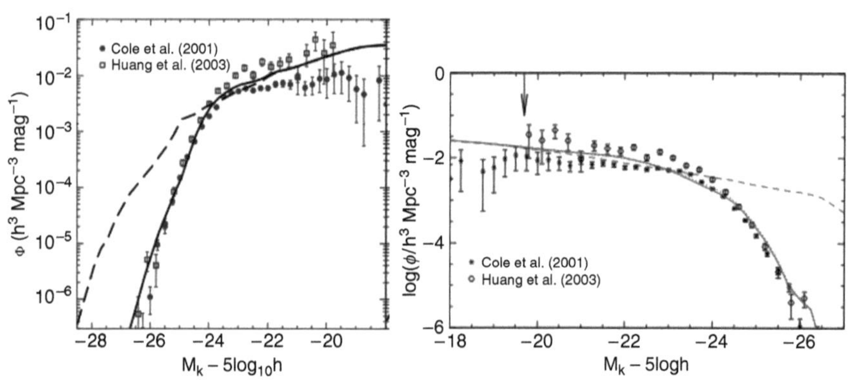

Fig. 3 *Left:* The predicted K-band luminosity function from two recently published semi-analytic models ([7] left panel and [2] *right panel*). In both panels, the *solid* and *dashed lines* show model predictions with and without AGN feedback, respectively

prescriptions adopted in these models are, however, necessarily very schematic and not well grounded in observation. A recent work by [12], in particular, has pointed out that the distributions of radio sources predicted by recent semi-analytic models is in disagreement with observational data, confirming that much work still needs to be done in order to understand exactly how and when AGN feedback is important.

4 What Next?

The above discussion highlights that there are several areas where our galaxy formation models could and should be improved. In particular:

- Different treatments of galaxy mergers lead to merger times that can differ up to one order of magnitude [9]. This has important consequences for the assembly history of the most massive galaxies, and for the evolution of the bright/massive end of the luminosity/mass function.
- Only recently, galaxy formation models have turned their attention to the formation of the diffuse stellar component. The increasing amount and quality of observational data on the intra-cluster light, and detailed comparisons with theoretical models, will provide important constraints on how this component forms and evolves as a function of cosmic time.
- Much recent work has focused on improving our treatment of gas stripping from satellite galaxies [14, 31]. These models, however, are not without problems: e.g. the model discussed in [14] still predicts a larger passive fraction among low-mass galaxies than is observed, and an excess of intermediate to low mass galaxies beyond $z \sim 0.5$. There results call for a deep revision of our modelling of the evolution of satellite galaxies.
- 'Radio-mode' AGN feedback represents an elegant solution to a number of long-standing problems related to the evolution of massive galaxies. Yet, our modelling of this physical process is still very schematic. In particular, the strong dependency on halo mass usually adopted in semi-analytic models might not be supported by observational data [12].

To conclude, I would like to remind that the importance of the physical processes discussed above has been investigated in detail using numerical simulations, at the typical velocity dispersions of galaxy clusters. Detailed studies at smaller scales (those of galaxy groups) are often lacking, even though groups likely represent the most common environment galaxies experience during their lifetimes.

The increasing amount and quality of the data being collected in these years are going to continuously challenge available and future models. This close link between theory and observations needs to be maintained (and strengthened) in order to improve our understanding of the processes driving galaxy formation and evolution as a function of the environment.

Acknowledgements I wish to thank A. Pasquali and I. Ferreras for the organization of a very lively and stimulating meeting. I acknowledge financial support from the European Research Council under the European Community's Seventh Framework Programme (FP7/2007-2013)/ERC grant agreement n. 202781.

References

1. J.E. Barnes, L. Hernquist, ApJ **471**, 115 (1996)
2. R.G. Bower et al., MNRAS **370**, 645 (2006)
3. M. Boylan-Kolchin, C.-P. Ma, E. Quataert, MNRAS **383**, 93 (2008)
4. M. Brüggen, G. De Lucia, MNRAS **383**, 1336 (2008)
5. L.L. Cowie, A. Songaila, Nature **266**, 501 (1977)
6. T.J. Cox et al., MNRAS **384**, 386 (2008)
7. D.J. Croton et al., MNRAS **365**, 11 (2006)
8. G. De Lucia et al., MNRAS **366**, 499 (2006)
9. G. De Lucia et al., MNRAS **406**, 1533 (2010)
10. R. Farouki, S.L. Shapiro, ApJ **243**, 32 (1981)
11. A.S. Font et al., MNRAS **389**, 1619 (2008)
12. F. Fontanot et al., MNRAS **413**, 957 (2011)
13. J.E. Gunn, J.R.I. Gott, ApJ **176**, 1 (1972)
14. Q. Guo et al., MNRAS **413**, 101 (2011)
15. B. Lanzoni et al., MNRAS **361**, 369 (2005)
16. R.B. Larson, B.M. Tinsley, C.N. Caldwell, ApJ **237**, 692 (1980)
17. C. Mastropietro et al., MNRAS **364**, 607 (2005)
18. I.G. McCarthy et al., MNRAS **383**, 593 (2008)
19. J.C. Mihos, in *Clusters of Galaxies: Probes of Cosmological Structure and Galaxy Evolution*, ed. by J.S. Mulchaey, A. Dressler, A. Oemler (Cambridge University press, London, 2004), p. 278
20. B. Moore, G. Lake, N. Katz, ApJ **495**, 139 (1998)
21. G. Murante et al., MNRAS **377**, 2 (2007)
22. P.E.J. Nulsen, MNRAS **198**, 1007 (1982)
23. T. Okamoto, M. Nagashima, ApJ **587**, 500 (2003)
24. D.O. Richstone, ApJ **204**, 642 (1976)
25. E. Roediger, M. Brüggen, MNRAS **380**, 1399 (2007)
26. A. Saro et al., MNRAS **406**, 729 (2010)
27. R.S. Somerville et al., MNRAS **391**, 481 (2008)
28. A. Toomre, J. Toomre, ApJ **178**, 623 (1972)
29. B. Vollmer et al., ApJ **561**, 708 (2001)
30. L. Wang et al., MNRAS **377**, 1419 (2007)
31. S.M. Weinmann et al., MNRAS **406**, 2249 (2010)
32. S.D.M. White, C.S. Frenk, ApJ **379**, 52 (1991)
33. S.D.M. White, M.J. Rees, MNRAS **183**, 341 (1978)

Reconciling a Significant Hierarchical Assembly of Massive Early-Type Galaxies at $z \lesssim 1$ with Mass Downsizing

M.C.C. Eliche-Moral, M. Prieto, J. Gallego, and J. Zamorano

Abstract Hierarchical models predict that massive early-type galaxies (mETGs) are the latest systems to be in place into the cosmic scenario (below $z \sim 0.5$), conflicting with the observational phenomenon of galaxy mass downsizing, which poses that the most massive galaxies have been in place earlier that their lower-mass counterparts (since $z \sim 0.7$). We have developed a semi-analytical model to test the feasibility of the major-merger origin hypothesis for mETGs, just accounting for the effects on galaxy evolution of the major mergers strictly reported by observations. The most striking model prediction is that very few present-day mETGs have been really in place since $z \sim 1$, because $\sim$90% of the mETGs existing at $z \sim 1$ are going to be involved in a major merger between $z \sim 1$ and the present. Accounting for this, the model derives an assembly redshift for mETGs in good agreement with hierarchical expectations, reproducing observational mass downsizing trends at the same time.

1 Assembly of mETGs in Groups and Field

Recent observations disagree with the late major-merger origin that hierarchical models predict for mETGs [2]. Massive galaxies ($\log(\mathcal{M}_*/\mathcal{M}_\odot) \gtrsim 11$, mostly E-S0a's) seem to have been already in place at $z \sim 0.7$, earlier than their less-massive counterparts (this phenomenon is known as mass downsizing, see [6]). Moreover, the S0 fraction in clusters increases with time at the expense of the spirals, being the fraction of ellipticals nearly constant [7]. This has been interpreted

M.C.C. Eliche-Moral (✉) · J. Gallego · J. Zamorano
Departamento de Astrofísica, Universidad Complutense de Madrid, 28040, Madrid, Spain
e-mail: mceliche@fis.ucm.es

M. Prieto
Instituto de Astrofísica de Canarias, C/Vía Láctea, 38200, La Laguna, Tenerife, Spain

I. Ferreras and A. Pasquali (eds.) *Environment and the Formation of Galaxies: 30 years later*, Astrophysics and Space Science Proceedings,
DOI 10.1007/978-3-642-20285-8_42,

as a sign of the poor role played by major mergers in the evolution of cluster ellipticals (contrary to the hierarchical picture). However, more than ~1/2 of present-day mETGs do not reside in clusters, but in groups [1], where mergers and tidal interactions determine the galaxy evolution [5]. So, the question is still open: are major mergers really the main drivers of the assembly of most mETGs? We have used semi-analytical models to test this hypothesis, studying how the present-day mETGs would have evolved backwards-in-time assuming that they derive exclusively from the (wet/mixed/dry) major mergers strictly reported by observations. The model predictions apply to field and group mETGs, as the local luminosity functions (LFs) and major merger fractions used here trace basically these environments [3, 4].

2 Reconciling Mass Downsizing and Hierarchical Scenarios

The model can reproduce the observed evolution of the galaxy LFs at $z \lesssim 1$, simultaneously for different rest-frame bands and selection criteria (based on color or morphology). The model predicts that ~90% of the mETGs existing at $z \sim 1$ are not the passively-evolved high-z counterparts of present mETGs (as usually interpreted in many studies), but their gas-poor progenitors instead.

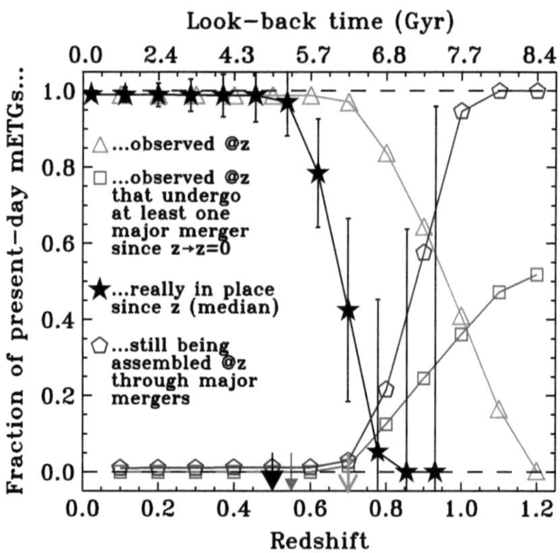

Fig. 1 Model predictions on the fraction of present-day mETGs observed, really in place, and still being assembled at each redshift z. *Arrows*: Assembly redshift of mETGs according to observations (*open*, [6]), to the model (*thick solid*), and average z between the assembly redshifts of 50% and 80% of their stellar content according to the models by [2] (*thin solid*). Our model reproduces the observed buildup of mETGs since $z \sim 1$ (triangles), predicting an assembly redshift for mETGs in good agreement with hierarchical models [2] at the same time

This implies that very few present-day mETGs have been really in place since $z \sim 1$ ($\lesssim 5\%$), contrary to the $\sim$50% reported by traditional interpretations of observations. Accounting for this, the model is capable of deriving a final assembly redshift of most mETGs in good agreement with hierarchical models ($z \sim 0.5$, neglecting minor mergers), reproducing global mass downsizing trends at the same time (see Fig. 1, adapted from [4]).

Acknowledgements Supported by the Spanish Ministry of Science and Innovation (MICINN) under project AYA2009-10368 and by the Madrid Regional Government through the AstroMadrid Project (CAM S2009/ESP-1496). Funded by the Spanish MICINN under the Consolider-Ingenio 2010 Program grant CSD2006-00070: "First Science with the GTC".

References

1. A.A. Berlind et al., ApJS **167**, 1 (2006)
2. G. De Lucia et al., MNRAS **366**, 499 (2006)
3. M.C. Eliche-Moral et al., A&A **519**, 55 (2010a)
4. M.C. Eliche-Moral et al., A&A (2010b, submitted). arXiv:1003.0686v1
5. S.J. Kautsch et al., ApJ **688**, L5 (2008)
6. P.G. Pérez-González et al., ApJ **687**, 50 (2008)
7. B.M. Poggianti et al., ApJ **697**, L137 (2009)

The Origin of the Morphology–Density Relation

D.J. Wilman, P. Erwin, G. De Lucia, F. Fontanot, and P. Monaco

Abstract The history of a galaxy is encoded in its morphology: the angular momentum and dissipative properties of gas leads to the formation of a disk, within which stars form; mergers of galaxies and tidal interactions randomize stellar orbits, forming a bulge; and the presence or lack of a gas disk drives the level of disk star formation and (apparently) spiral features. The history of a galaxy is also encoded in its environment: sub-halos and their galaxies rarely merge, except at the bottom of the potential well (the centre of the main halo) in which they live; and gas disks and the hot gas which feeds them can only be stripped in a dense surrounding medium, found only in massive haloes (e.g. [Chung et al., ApJ 138, 1741 (2009), McCarthy et al., MNRAS 383, 593 (2008)]). To distinguish the mechanisms driving galaxy evolution, we examine the halo mass dependence of morphology separately for central and satellite galaxies at $z \sim 0$. We place constraints on evolution utilizing a variety of physical prescriptions applied to semi-analytic models and compared to a SDSS-RC3 matched sample at $z \sim 0$. We find ellipticals primarily at the centre of $M_{halo} \gtrsim 10^{13}\,M_{\odot}$ haloes, compatible with expectations of bulge growth in mergers; S0s and passive disk galaxies are created via two independent channels: as central galaxies in $M_{halo} \gtrsim 10^{12}\,M_{\odot}$ haloes, with star formation suppressed possibly as a result of AGN feedback, and as satellite galaxies in $M_{halo} \gtrsim 10^{13}\,M_{\odot}$ haloes in which stripping of the hot gas associated to a galaxy is probably responsible for the suppression.

D.J. Wilman (✉) · P. Erwin
Max-Planck-Institut für extraterrestrische Physik, Giessenbachstraße, 85748 Garching, Germany
e-mail: dwilman@mpe.mpg.de; erwin@mpe.mpg.de

G. De Lucia · F. Fontanot · P. Monaco
Osservatorio Astronomico di Trieste, Via Tiepolo 11 34143 Trieste, Italy
e-mail: delucia@oats.inaf.it; fontanot@oats.inaf.it; monaco@oats.inaf.it

I. Ferreras and A. Pasquali (eds.) *Environment and the Formation of Galaxies: 30 years later*, Astrophysics and Space Science Proceedings,
DOI 10.1007/978-3-642-20285-8_43, © Springer-Verlag Berlin Heidelberg 2011

1 A Brief Background

The discovery of a tight relation between the fraction of galaxies with a given morphology and their environment puts strong constraints on the physical processes driving galaxy evolution [5]. Elliptical galaxies are typically found in regions of very high density, or in the cores of haloes, but with a global luminosity-limited fraction which is otherwise almost independent of environment and redshift at $z \lesssim 1$ [12]. Lenticular (S0) galaxies are found in similar numbers in group and cluster haloes, increasing dramatically since $z \sim 0.5$ [6, 7], at least at $\sigma_{clus} \lesssim 750\,\mathrm{km\,s^{-1}}$ [10], but are less frequent in the lower density field [12]. In this contribution, we describe ongoing work utilizing both observations and models to probe the physical processes driving the co-evolution of galaxy morphology and star formation as a function of environment.

2 The SDSS-RC3 Catalogue

To examine the local relationship between galaxy morphology and halo mass, we take the morphological classifications from the RC3 catalogue of galaxies [4], and match this sample to the Sloan Digital Sky Survey (SDSS) Data Release 4 (DR4) [1]. The RC3 catalogue selection function is quantified (by comparing the matched sample to an underlying parent sample selected from SDSS) as a function of b-magnitude[1] and r_{50}.[2] Galaxies are then weighted by the inverse of the selection function, and limited to $b \leq 16.0$. B-band luminosities are computed using the *kcorrect* code of [2]. Morphological fractions are computed down to a limiting magnitude of $\mathrm{M_B} = -19$, and weighted by V/V_{max} where V_{max} is the volume within which a given galaxy is visible.

Halo masses and the central (most massive in the group) or satellite galaxy status are taken from the SDSS-DR4 based group catalog of [14], which is complete at $\mathrm{M_{halo}} \geq 10^{11.6348}\,\mathrm{M_\odot}$.

3 The Models

Our models are based upon the semi-analytical model of [11], itself based upon the model described by [3]. Each model prescription is applied to a dark matter halo catalogue based on a N-body simulation in a ΛCDM Universe. The assembly of each halo, subhalo and galaxy is tracked via merger-trees, keeping an inventory

---

[1] $b = u - 0.8116(u - g) + 0.1313$ (http://www.sdss.org/dr7/algorithms/sdssUBVRITransform.html #Lupton2005).

[2] The radius containing 50% of the Petrosian flux in r-band.

of the properties of each galaxy including stellar mass, star formation history and bulge to total ratio (B/T). As we only consider the morphology-altering B/T and star formation of galaxies, the most relavent physical processes can be summarized as follows:

- *Galaxy mergers:* Bulge growth is mainly driven by the redistribution and randomization of stellar orbits during a major merger of disk galaxies. We consider two prescriptions for bulge growth during a merger:

  1. The default prescription [3] in which the final bulge consists of all stars from the secondary (less massive) progenitor and in the case of a major (mass ratio, $\mu > 0.3$) merger, all stars from the primary progenitor. A fraction of the cold gas $f_{SB} = 0.56.\mu^{0.7}$ is converted to stars during the merger via a starburst, which deposits new stars *in the disk.*
  2. Our second prescription (the *Hopkins* prescription) is based on the results of recent hydro-dynamical simulations of [8, 9]. This prescription differs from the default in two ways. The fraction of stars from the primary galaxy which are deposited in the final bulge is simply $\mu\,\mathrm{M}_*^{\mathrm{primary}}$. The fraction of cold gas which participates in the starburst is a decreasing function of initial gas fraction and an increasing function of mass ratio. In this case, the starburst deposits the new stars *in the bulge*.

- *Disk instabilities:* We apply our default model twice, with the second channel for bulge formation, related to secular disk instabilities, switched on and off (to investigate the effect of mergers in isolation). The *Hopkins* prescription is applied with disk instabilities switched off.
- *Star formation, cooling and heating:* The availability of gas for star formation depends to first order on the balance of hot and cold gas belonging to the galaxy. Gas is typically shock-heated upon accretion onto the halo, subsequently cooling onto the galaxy, forming stars in a disk, and also providing energy via supernova feedback which reheats the gas. Gas is also reheated and ejected by radio-mode AGN in the centre of massive haloes.
- *Stripping of hot gas:* Hot gas is stripped instantaneously and entirely upon the accretion of the galaxy onto a more massive halo (the galaxy becomes a satellite). Cooling is then suppressed, and the remaining cold gas is used up via star formation and heating.

4 Results

Figure 1 (top panels) shows how the fraction of $\mathrm{M_B} \leq -19$ central (left) and satellite (right) galaxies with elliptical morphology (or $\mathrm{B/T} \geq 0.9$ in the models) depends on halo mass. The points show the weighted fraction in the SDSS-RC3 sample, with 68% confidence limits computed using the Wilson method to estimate binomial errors [13], first rescaling all (weighted) counts so that the total counts are equal to

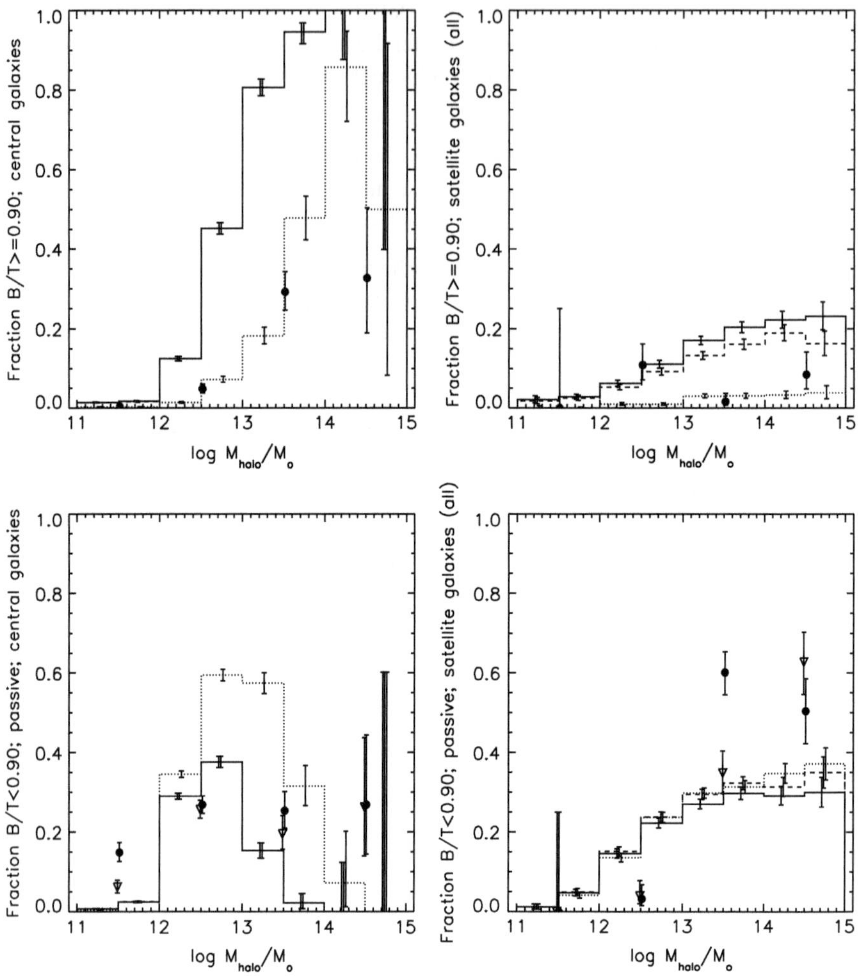

Fig. 1 *Above*: Fraction of SDSS-RC3 Elliptical galaxies (*points with errors*) and fraction of model galaxies with $B/T \geq 0.9$ (*solid/dashed/dotted lines* = merger+instability/pure merger/Hopkins recipe respectively) as a function of halo mass, separately for central (*left*) and satellite (*right*) galaxies. *Below*: Fraction of SDSS-RC3 S0s (*circles with errors*) and passive-core disk galaxies (*open triangles with errors*), and fraction of model galaxies with $B/T < 0.9$ and no current star formation (*lines*, as above) as a function of halo mass, separately for central (*left*) and satellite (*right*) galaxies

the original (unweighted) total counts. The lines illustrate the equivalent fractions (and confidence limits) computed at $z = 0$ for the three model prescriptions.

Elliptical galaxies are preferentially found at the centre of $M_{halo} \gtrsim 10^{13}\,M_{\odot}$ haloes, and this is consistent with the models. Galaxy mergers take place at the centre of haloes, and thus the galaxies which live in this environment, especially those in the most massive haloes, have a richer merger history with associated bulge growth. In the highest mass haloes the formation of a new disk is also suppressed by AGN feedback. Disk instabilities are almost irrelevant to the fraction and environmental dependence of ellipticals in our models, whilst the Hopkins recipe appears to provide a better match to the low fraction of ellipticals, since a higher fraction of cold gas can survive at high initial gas fractions, providing the raw material for a new stellar disk.

Figure 1 (lower panels) shows the fraction of $M_B \leq -19$ central (left) and satellite (right) galaxies with S0 morphology in SDSS-RC3 (circles with errors) or S0/spiral morphology and insignificant Hα emission in the SDSS fiber (passive-core, open triangles with errors) as a function of halo mass. Overplotted are the fraction of B/T < 0.9 galaxies with no star formation (SFR $= 0$) in the models (lines). Two important points can be taken from this figure:

- The observations illustrate that there are two separate populations of S0s and passive-core spirals: one in central galaxies, ~25% of galaxies in $M_{halo} \gtrsim 10^{12}\,M_\odot$ haloes, and one in satellite galaxies, but only in $M_{halo} \gtrsim 10^{13}\,M_\odot$ haloes. The suppression of spiral arms and central star formation is likely to relate to different mechanisms in these two sub-populations of disk galaxies.
- Model suppression of star formation in satellite galaxies can qualitatively reproduce the halo mass dependence of passive disks, although more work is needed to understand this quantitatively. Suppression by AGN of star formation in central galaxies is possibly too efficient: the balance between suppression of star formation and suppression of disks produces a peak at $M_{halo} \sim 10^{13}\,M_\odot$ which is not observed.

5 Further Comments

Whilst the environmental dependence of the elliptical fraction at $z = 0$ can be easily understood in the context of hierarchical models, the lack of evolution to $z \sim 1$ in clusters is not reproduced: the fraction of high B/T galaxies in all models is increasing, in all environments. In fact, most ellipticals in the model *regrow their disks*, even in the default model for which the growth is mainly driven by a cooling flow onto the galaxy. The strength of this regrowth rate and its influence on morphology at later times are a key aspect of theoretical modeling that we will fully address in following work. This work paves the way to a more detailed understanding of the relationships between morphology, star formation and AGN activity and environment and thence the physical mechanisms driving these observed relationships.

References

1. J.K. Adelman-McCarthy et al., ApJS **162**, 38 (2006)
2. M.R. Blanton, S. Roweis, AJ **133**, 734 (2007)

3. G. De Lucia, J. Blaizot, MNRAS **375**, 2 (2007)
4. G. de Vaucouleurs et al., *Third Reference Catalogue of Bright Galaxies*, Springer-Verlag (1991)
5. A. Dressler, ApJ **236**, 351 (1980)
6. A. Dressler et al., ApJ **490**, 577 (1997)
7. G. Fasano et al., ApJ **542**, 673 (2000)
8. P.F. Hopkins et al., ApJ **691**, 1168 (2009)
9. P.F. Hopkins et al., MNRAS **397**, 802 (2009)
10. D.W. Just et al., ApJ **711**, 192 (2010)
11. J. Wang et al., MNRAS **384**, 1301 (2008)
12. D.J. Wilman et al., ApJ **692**, 298 (2009)
13. E.B. Wilson, J. Am. Statist. Assoc. **22**, 209 (1927)
14. X. Yang et al., ApJ **671**, 153 (2007)

Towards Understanding Simulations of Galaxy Formation

N.L. Mitchell

Abstract Numerical simulations are now a fundamental tool with which modern astronomers test current theory. However an increasing number of authors have noted significant discrepancies between galaxy properties when run in different hydrodynamic codes. As we can now finally begin to run large cosmological simulations with complex gas physics it is necessary to understand the way in which these differences between codes affect the properties of the ISM. From the efficiency of supernova feedback to the large scale heating of gas during galaxy cluster mergers, I will show that there are notable differences between particle and grid based codes, explain their origin and demonstrate work that is being undertaken at Vienna to improve the way in which we model the properties of the ISM.

1 The Origin of Entropy Cores in Simulated Galaxy Clusters

The Santa Barbara cluster comparison showed that Adaptive Mesh Refinement codes (AMR) produce hotter, higher entropy cores than in Smooth Particle Hydrodynamics codes (SPH). Despite many groups attempting to determine the underlying cause of this discrepancy through the use of large suites of cosmological simulations, their complexity prevented the true cause from being isolated. In [1] we simplify the problem through the use of mergers between idealised model clusters. We conclude that the spurious suppression of hydrodynamic instabilities in SPH codes leads to the absence of turbulent mixing present in AMR codes. This turbulent mixing allows higher entropy material on the outside of the cluster to mix with lower entropy material within the core, whereas the low entropy material is undisturbed in

N.L. Mitchell (✉)
Institute für Astronomie, Türkenschanzstrasse, 1180 Vienna
e-mail: nigel.mitchell@univie.ac.at

I. Ferreras and A. Pasquali (eds.) *Environment and the Formation of Galaxies: 30 years later*, Astrophysics and Space Science Proceedings,
DOI 10.1007/978-3-642-20285-8_44, © Springer-Verlag Berlin Heidelberg 2011

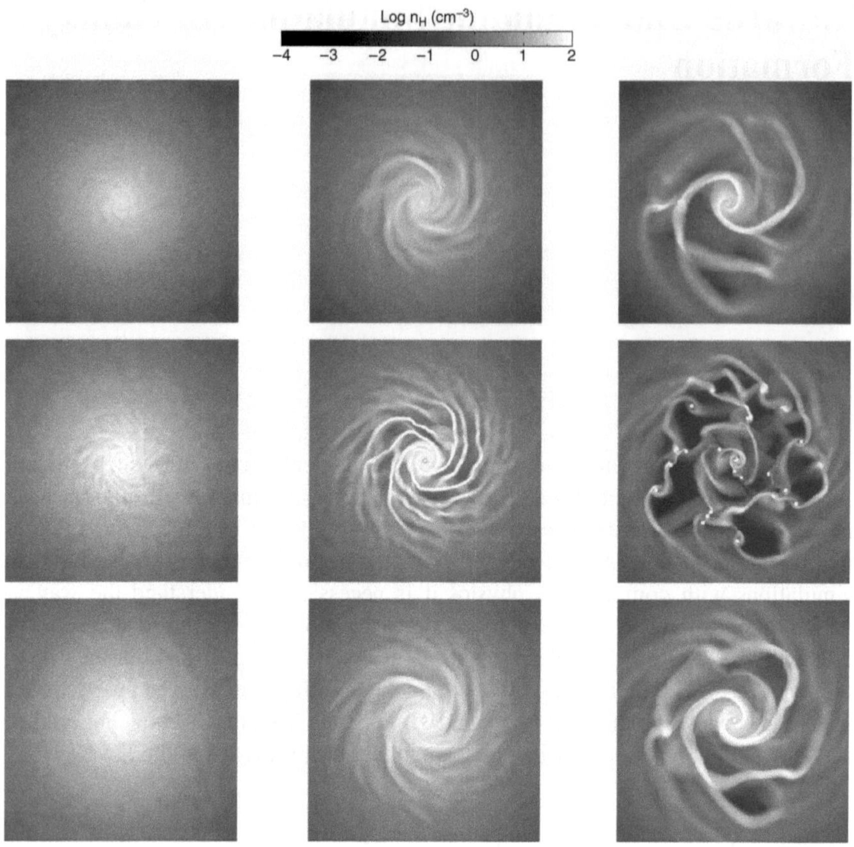

Fig. 1 *Top down* images of spiral galaxies run with SPH (*top panel*: $\gamma_{eff} = 4/3$) and AMR (*centre*: $\gamma_{eff} = 4/3$, *bottom*: $\gamma_{eff} = 5/3$)

SPH codes. Such a difference can allow AMR codes to more easily form non-cool core clusters without the need for as much energetic feedback.

2 How Sub-Grid Models for Star Formation are Affected by the Choice of Hydrodynamic Implementation

Many galaxy simulations which lack the resolution to resolve the multi-phase nature of the ISM use polytropic equations of state to approximate the pressure, P,

$$P \propto \rho_g^{\gamma_{eff}}, \tag{1}$$

where ρ_g is the gas density and γ_{eff} is the polytropic index. This index is usually chosen in SPH codes such that the smoothing length remains a fixed fraction of the Jeans length. To determine how hydrodynamic implementations affect the underlying sub-grid physics I implement the star formation and cooling routines used in [2, 3] into the AMR code FLASH. When disk galaxy simulations are run in AMR and SPH codes using identical pressure laws, the galaxy dynamics differ dramatically as can be seen in Fig. 1. When the spiral arms wind up and touch, in the SPH code (top) they remain stable whilst in the AMR code (centre) they collapse down into dense knots which then randomly collide, completely disrupting the spiral arms. By increasing the polytropic index in the AMR code to $\gamma = 5/3$ (bottom), the pressure support against collapse and thus effective Jeans length is increased, suppressing the formation of dense knots. This yields almost perfectly identical spiral arms to those in the SPH code. Thus identical sub-grid recipes will not necessarily yield identical results.

References

1. N.L. Mitchell et al., MNRAS **395**, 180M (2009)
2. J. Schaye, C. Dalla Vecchia, MNRAS **383**, 1210–1222 (2008)
3. R.P.C. Wiersma, J. Schaye, B.D. Smith, MNRAS **393**, 99 (2009)

Quadruple-Peaked Line-of-Sight Velocity Distributions in Shell Galaxies

I. Ebrová, L. Jílková, B. Jungwiert, K. Bartošková, M. Křížek, T. Bartáková, and I. Stoklasová

Abstract We present an improved study of the expected shape of the line-of-sight velocity distribution in shell galaxies. We found a simple analytical expression connecting prominent and in principle observable characteristics of the line profile and mass-distribution of the galaxy. The prediction was compared with the results from a test-particle simulation of a radial merger.

1 Quadruple-Peaked Spectral-Line Profile

Stellar shells are observed in almost half of elliptical and S0 galaxies that live in a low galactic density environment, see e.g. [1]. They are thought to be by-products of galaxy mergers [5]. The most regular shell systems are believed to result from a nearly minor radial merger in which the satellite galaxy is dissolved by tidal forces and its stars begin to oscillate in the potential of the host galaxy at close-to-radial orbits. The stars accumulate at their turning points and create shells.

I. Ebrová (✉) · B. Jungwiert · M. Křížek
Astronomical Institute, Academy of Sciences of the Czech Republic
and
Faculty of Mathematics and Physics, Charles University in Prague
e-mail: ivana@ig.cas.cz

L. Jílková
ESO Santiago, Chile
and
Faculty of Science, Masaryk University, Brno, Czech Republic

K. Bartošková and T. Bartáková
Faculty of Science, Masaryk University, Brno, Czech Republic

I. Stoklasová
Astronomical Institute, Academy of Sciences of the Czech Republic

I. Ferreras and A. Pasquali (eds.) *Environment and the Formation of Galaxies: 30 years later*, Astrophysics and Space Science Proceedings,
DOI 10.1007/978-3-642-20285-8_45, © Springer-Verlag Berlin Heidelberg 2011

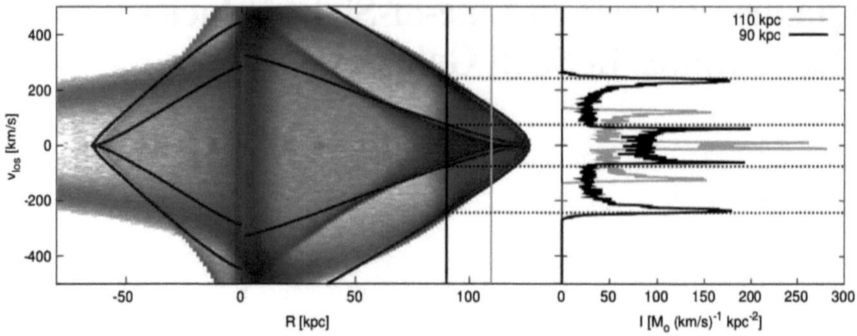

Fig. 1 *left:* LOSVD map of the simulated shell galaxy (only stars of the satellite galaxy are taken into account). The two apexes of the wedges seen in the map (at zero velocity) correspond to the two shells, the *black curves* show the velocity maxima position obtained from our approximation. *right:* LOSVDs (cuts of the map shown in the *left* plot) of stars belonging to the right shell at two different galactocentric distances – 90 and 110 kpc – *black* and *gray* profiles, respectively. *Dashed lines* show the locations of the maxima in our approximation for the *black* (90 kpc) profile

The shape of the line-of-sight velocity distribution (LOSVD) in the vicinity of the shell edge for a stationary shell was studied by [4]. They predicted a double-peaked spectral-line profile and proposed to use spectroscopy to probe the dark matter distribution of a galaxy that contains shells using the profiles of stellar absorption lines.

Nevertheless, shells are not stationary features: stars of the satellite galaxy have a continuous energy distribution, and therefore the shell edge is, at different times, made of stars of different energies, as they continue to arrive at their respective turning points. Thus, the shell front moves outwards from the center of the host galaxy with its velocity given by the mass distribution of the host galaxy. Therefore, both of the original double peaks in the spectral line are split into two, resulting in a *quadruple-peaked shape* [3]. Taking the shell's velocity and the cumulative mass of the host galaxy to be constant near the edge of the shell, we found an approximate analytical description for the positions of the peaks in the LOSVD (for details see [2]).

To study the LOSVD more in detail we carried out a test-particle simulation of a radial merger of dwarf (dE) and giant elliptical (gE) galaxies, leading to a formation of shells. The potential of the gE galaxy is represented with a luminous de Vaucouleurs sphere and an NFW dark halo. See Fig. 1 for comparison of LOSVD from simulation and the analytical approximation. If the velocity maxima were measured, the approximation could be used to constrain the mass distribution of the host galaxy.

Acknowledgements We acknowledge the support by the grant No. 205/08/H005 (Czech Science Foundation), the project SVV 261301 (Charles University in Prague), EAS grant covering registration fee at JENAM 2010, the grant MUNI/A/0968/2009 (Masaryk University in Brno), the grant LC06014, Center for Theoretical Astrophysics (Czech Ministry of Education) and research plan AV0Z10030501 (Academy of Sciences of the Czech Republic).

References

1. J.W. Colbert et al., AJ **121**, 808 (2001)
2. I. Ebrová et al., (2011, in preparation)
3. L. Jílková et al., ASCP **423**, 243 (2010)
4. M.R. Merrifield, K. Kuijken, MNRAS **297**, 1292 (1998)
5. P.J. Quinn, ApJ **279**, 596 (1984)

Tidal Stirring of Milky Way Satellites: A Simple Picture with the Integrated Tidal Force

E.L. Łokas, S. Kazantzidis, L. Mayer, and S. Callegari

Abstract Most of dwarf spheroidal galaxies in the Local Group were probably formed via environmental processes like the tidal interaction with the Milky Way. We study this process via N-body simulations of dwarf galaxies evolving on seven different orbits around the Galaxy. The dwarf galaxy is initially composed of a rotating stellar disk and a dark matter halo. Due to the action of tidal forces it loses mass and the disk gradually transforms into a spheroid while stellar motions become increasingly random. We measure the characteristic scale-length of the dwarf, its maximum circular velocity, mass, shape and kinematics as a function of the integrated tidal force along the orbit. The final properties of the evolved dwarfs are remarkably similar if the total tidal force they experienced was the same, independently of the actual size and eccentricity of the orbit.

1 Introduction

In the tidal stirring scenario for the formation of dwarf spheroidal (dSph) galaxies [9] the progenitors are late type dwarfs affected by tidal forces from the Milky Way or any other normal size galaxy. A similar process may lead to the formation of S0s

E.L. Łokas (✉)
Nicolaus Copernicus Astronomical Center, 00-716 Warsaw, Poland
e-mail: lokas@camk.edu.pl

S. Kazantzidis
Center for Cosmology and Astro-Particle Physics; and Department of Physics; and Department of Astronomy, The Ohio State University, Columbus, OH 43210, USA
e-mail: stelios@mps.ohio-state.edu

L. Mayer · S. Callegari
Institute for Theoretical Physics, University of Zürich, 8057 Zürich, Switzerland
e-mail: lucio@phys.ethz.ch; callegar@physik.uzh.ch

I. Ferreras and A. Pasquali (eds.) *Environment and the Formation of Galaxies: 30 years later*, Astrophysics and Space Science Proceedings,
DOI 10.1007/978-3-642-20285-8_46, © Springer-Verlag Berlin Heidelberg 2011

in galaxy clusters [2,3]. In the absence of any analytical description of tidal effects for dwarfs on eccentric orbits, that seem to dominate in ΛCDM cosmologies, the problem can be fully addressed only via N-body simulations. With this tool, the tidal stirring scenario has been recently tested for a variety of orbits and structural parameters of the dwarfs and shown to work very efficiently towards the formation of dSph galaxies [4–8]. It was demonstrated that during the tidal evolution the dwarfs lose mass via tidal stripping, they undergo morphological transformation from disks to spheroids and the rotation of their stars is replaced by random motions.

However, the exact mechanism behind this transformation still eludes us. It is generally believed that the processes shaping the dwarf galaxies orbiting on eccentric orbits may be inherently different from those on circular orbits due to the time-dependence of the tidal force. While on circular orbits tidal forces are believed to mainly steepen the outer density profile, on eccentric orbits they likely induce strong shocks to the whole structure of the dwarf galaxy at pericenters [1]. It has also been suggested, however, that the key factor that controls the extent of transformation is the integrated tidal force the dwarf experiences along the orbit rather than the particular shape of the orbit [9]. Here we address this question using a subset of collisionless N-body simulations described in detail in [4].

In these numerical experiments a dwarf galaxy, initially composed of a stellar disk and a dark matter halo, is placed on seven different orbits O1–O7 around a live Milky Way model. The orbital parameters of the simulations are listed in Table 1. Orbits O1–O5 correspond to runs R1–R5 in [4], while orbits O6 and O7 are two additional setups. The second and third column of the table list the apo- and pericenter distances and the fourth one the orbital time. All simulations were evolved for 10 Gyr; the fifth column gives the time when the last apocenter occurred and the last column the total number of apocenters.

Our dwarf galaxy model was identical in all runs (model D1 in [4]). It contained a stellar disk of mass $M_{\rm d} = 2 \times 10^7\ {\rm M}_\odot$ with scale-length $R_{\rm d} = 0.41$ kpc and thickness parameter $R_{\rm d}/z_{\rm d} = 0.2$. The dwarf's dark matter halo had an NFW profile with virial mass $M_{\rm h} = 10^9\ {\rm M}_\odot$ and concentration $c = 20$. The mass distribution of the Milky Way was given by model MWb in [10].

Table 1 Orbital parameters of the simulated dwarfs

Orbit	$r_{\rm apo}$ [kpc]	$r_{\rm peri}$ [kpc]	$T_{\rm orb}$ [Gyr]	$t_{\rm la}$ [Gyr]	$n_{\rm apo}$
O1	125	25	2.09	8.35	5
O2	87	17	1.28	8.95	8
O3	250	50	5.40	5.40	2
O4	125	12.5	1.81	9.05	6
O5	125	50	2.50	10.00	5
O6	80	50	1.70	8.50	6
O7	250	12.5	4.55	9.10	3

2 Results

For each of the simulations we calculated the evolution of a few key properties of the dwarf in time: the characteristic radius $r_{\max}$ at which the circular velocity has a maximum, the maximum circular velocity $V_{\max}$, and the mass contained within the characteristic scale, $M(<r_{\max})$. Using stars inside $r < r_{\max}$ we determined the shape and kinematic properties of the stellar component. The shape is described in terms of the shortest to longest axis ratio c/a calculated from the moments of the inertia tensor. The amount of ordered versus random motion was quantified by the ratio of the rotation velocity around the shortest axis to the 1D velocity dispersion, V/σ. The anisotropy of the stellar motions was characterized by the usual β parameter. The evolution of these properties as a function of time is illustrated in Fig. 1 for the least and the most eccentric orbit (O6 and O7). The results for other orbits are discussed in [4].

Next, we calculated the tidal force experienced by the dwarf galaxy on different orbits as

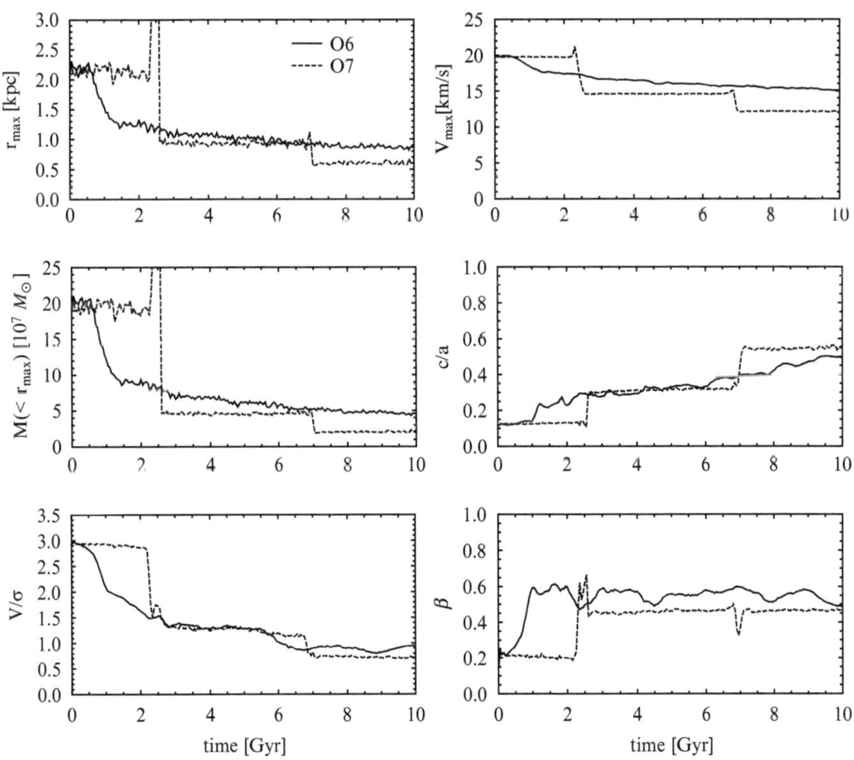

Fig. 1 Evolution of the properties of the dwarf as a function of time for the least (O6, *solid line*) and the most (O7, *dashed line*) eccentric orbit

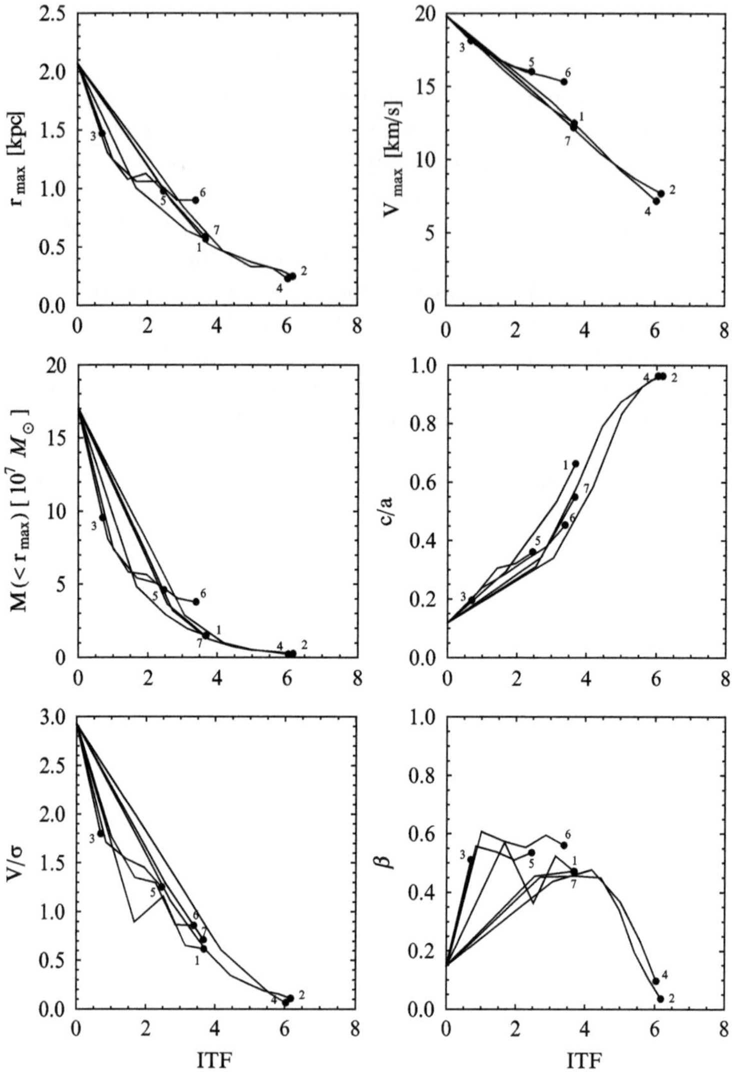

Fig. 2 Evolution of the properties of the dwarf as a function of the integrated tidal force (ITF, in arbitrary units). The *lines* connect measurements at subsequent apocenters for a given orbit with *dots* marking the results at the last apocenter. In each panel all lines start from the same point at ITF = 0 because the initial structural properties of the dwarf were the same for all orbits and all simulations started at apocenters

$$F_{\rm tidal} \propto r_{\rm max} M(< R)/R^3 \ , \tag{1}$$

where $r_{\rm max}$ is the characteristic scale-length of the dwarf at which the tidal force operates, R is the distance between the center of the dwarf and the center of the Milky Way and $M(< R)$ is the mass of the Milky Way model within this distance

Table 2 Properties of the simulated dwarfs at the last apocenter

Orbit	r_{max} [kpc]	V_{max} [km/s]	$M(< r_{max})$ [10^7 $M_\odot$]	c/a	V/σ	β
O1	0.57	12.5	1.50	0.66	0.62	0.47
O2	0.25	7.7	0.27	0.96	0.11	0.04
O3	1.47	18.1	9.56	0.20	1.80	0.51
O4	0.23	7.2	0.22	0.96	0.07	0.10
O5	0.98	16.0	4.61	0.36	1.25	0.54
O6	0.90	15.3	3.79	0.45	0.86	0.56
O7	0.59	12.2	1.48	0.55	0.71	0.47

(approximated as a spherical NFW distribution with the disk and the bulge added as point masses). Summing up contributions of this form over a given orbit we obtain an estimate of the integrated tidal force (ITF) experienced by the dwarf up to a given time. To make the results comparable for different orbits it is advisable to integrate over full orbital times so for each of the orbits we summed up the tidal force from the first up to the last apocenter.

The evolution of different properties of the dwarf, described above, as a function of the integrated tidal force, is shown in Fig. 2. The lines join the results for a given orbit at subsequent apocenters with the dots marking the last apocenter. Numbers near the points indicate the orbit, with 1–7 corresponding to orbits O1–O7 in Table 1. The values of the parameters of the dwarf at the last apocenter are listed in Table 2.

3 Conclusions

The trends of the dwarf galaxy properties as a function of the integrated tidal force are remarkably similar. In spite of different size and eccentricity of the orbits the evolution seems to be controlled almost entirely by the amount of tidal force experienced by the dwarf. The similarity is particularly striking in the case of two pairs of orbits, O2–O4 and O1–O7. Although the orbital parameters in each pair are very different (see Table 1) the dwarfs experience a very similar integrated tidal force and all their properties at the last apocenter are also almost identical. A slight departure from the trend set by the orbits O2–O4 and O1–O7 is seen only in the case of V_{max} for orbits O5–O6 with the largest pericenter.

Acknowledgements This research and ELŁ's participation in JENAM 2010 was partially supported by the Polish Ministry of Science and Higher Education under grant NN203025333.

References

1. J. Binney, S. Tremaine, *Galactic Dynamics* (Princeton University Press, Princeton, 2008)
2. O.Y. Gnedin, ApJ **582**, 141 (2003)

3. O.Y. Gnedin, ApJ **589**, 752 (2003)
4. S. Kazantzidis et al., ApJ **726**, 98 (2011)
5. J. Klimentowski et al., MNRAS **397**, 2015 (2009)
6. E.L. Łokas et al., ApJ **708**, 1032 (2010)
7. E.L. Łokas et al., ApJ **725**, 1516 (2010)
8. Mayer et al., Nature **445**, 738 (2007)
9. L. Mayer et al., ApJ **559**, 754 (2001)
10. L.M. Widrow, J. Dubinski, ApJ **631**, 838 (2005)

Ram Pressure Stripping of Hot Galactic Halos in Galaxy Clusters

V. Baumgartner and D. Breitschwerdt

Abstract The intracluster medium (ICM) in galaxy clusters contains heavy elements with about 1/3 of the solar abundance. These heavy elements are the products of stellar nucleosynthesis and are either expelled by galactic winds or lost from the galaxies due to interactions with the intracluster gas. We investigate the stripping of hot, high-metallicity galactic halos, which occurs as galaxies are moving through a cluster, being subject to the ram pressure of the ICM. Our new model for ram pressure stripping differs from earlier models, since it includes processes inside the galaxies like the transport of gas from the disk into the halo. Taking into account different components of the interstellar medium and their vertical distribution, we get a more realistic and detailed stripping criterion.

1 Metal Enrichment of the ICM

X-ray observations reveal that the ICM filling the space among the galaxies in a galaxy cluster is enriched with heavy elements like iron, silicon and oxygen (metals). Since these elements are exclusively produced by stars inside the cluster galaxies, they must have been transported into intracluster space. Several mechanisms are known so far for the metal enrichment of the ICM which we have to investigate in order to understand the chemical composition of galaxies and clusters as well as feedback processes on stellar evolution. While galactic winds expel processed gas out of the galaxies into the ICM, the ram pressure of the

V. Baumgartner (✉)
Institut f. Astronomie, Universität Wien, Türkenschanzstraße 17, 1180 Vienna, Austria
e-mail: verena.baumgartner@univie.ac.at

D. Breitschwerdt
Zentrum f. Astronomie und Astrophysik, Technische Universität Berlin, Hardenbergstraße 38, 10623 Berlin, Germany

I. Ferreras and A. Pasquali (eds.) *Environment and the Formation of Galaxies: 30 years later*, Astrophysics and Space Science Proceedings,
DOI 10.1007/978-3-642-20285-8_47, © Springer-Verlag Berlin Heidelberg 2011

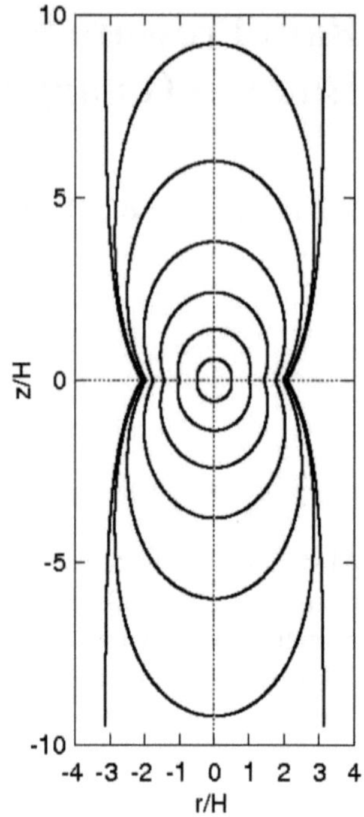

Fig. 1 Evolution of a superbubble in an exponentially stratified medium symmetric to the galactic midplane. The position of the shock front at different evolutionary stages is shown. The radius r and the height z above the plane are given in scale height units of the ISM

ICM leads to stripping of the interstellar medium (ISM) while a galaxy is moving through a cluster. In contrast to the stripping of galactic disks, which is efficient in the cluster center, halo stripping may occur in every region of a cluster, thus being a very important enrichment process. We investigate the stripping of hot, high-metallicity galactic halos and present the first attempts to a new model for ram pressure stripping which includes the matter distribution of the gaseous halo.

2 Disk-Halo-Connection

Previous stripping criteria apply to a thin, homogeneous disk described by its surface density only [1], or implement the gravitational acceleration of bulge, dark matter halo and galactic disk [3], neglecting the component of a hot, gaseous halo. In order to develop a new model, we start with analyzing in detail the gas transport from the galactic disk into the halo and the dynamic structure of the halo. Based on the approximation by Kompaneets [4], the expansion of superbubbles (SBs) – formed by sequential type II supernova explosions in the disk – in an exponentially stratified medium [2] is described (see Fig. 1). Using a stellar initial mass function and the main sequence life time of massive stars we get the time-dependent evolution of SBs. We follow the acceleration of SBs into the halo until they fragment and calculate the density and metal content of supernova ejecta inside the SBs. This knowledge is the basis for the description of the hot interstellar medium beyond the point where SBs fragment.

Acknowledgements VB is recipient of a DOC-fFORTE fellowship of the Austrian Academy of Sciences.

References

1. M.G. Abadi, B. Moore, R.G. Bower, MNRAS **308**, 947 (1999)
2. V. Baumgartner, D. Breitschwerdt (in preparation)
3. J.E. Gunn, J.R. Gott III, ApJ **176**, 1 (1972)
4. A.S. Kompaneets, Sov. Phys. Dokl. **5**, 46 (1960)

The Transformation of Virgo Galaxies Under the Influence of Ram Pressure

S. Boissier and A. Boselli

Abstract Simple models of chemical and spectro-photometric evolution of galaxies allow to study the transformation of galaxies under the influence of e.g. ram pressure in the Virgo cluster (other interactions can also be modeled). With the new data of the large surveys of Virgo in progress (e.g. NGVS in the optical, GUViCS in the UV), we will be able to extend this type of analysis to a large number of galaxies and get a better understanding of the transformation of galaxies in the high density environment.

1 The Effect of Ram Pressure Events on Large Spirals in Virgo

Ram pressure affects the photometric profiles of galaxies, with different effects at various wavelengths, depending on the age of the event. The multi-wavelength profiles of NGC4569 allowed [4] to date and measure the efficiency of a ram pressure event by including its effect in the grid of reference models "without interaction" of [1]. UV and optical profiles are of paramount importance for obtaining such results, as the ram-pressure produces a reversal of some of the color gradients, depending on the age of the interaction.

S. Boissier (✉) · A. Boselli
Laboratoire d'Astrophysique de Marseille, UMR6110, CNRS/Université de Provence, 38, rue Frdric Joliot-Curie, 13388 Marseille cedex 13, France
e-mail: boissier@oamp.fr; boselli@oamp.fr

I. Ferreras and A. Pasquali (eds.) *Environment and the Formation of Galaxies: 30 years later*, Astrophysics and Space Science Proceedings,
DOI 10.1007/978-3-642-20285-8_48,

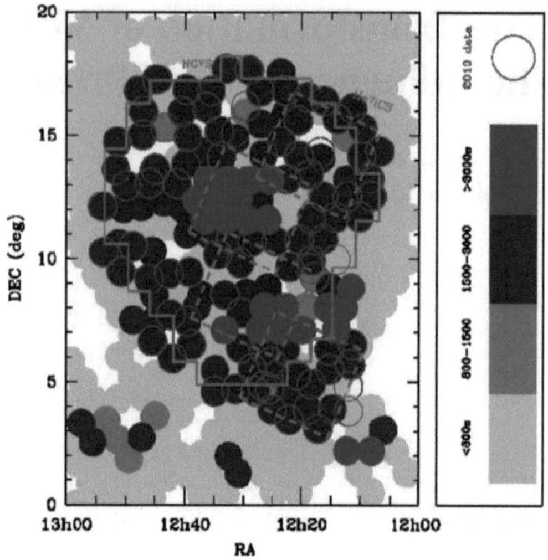

Fig. 1 Coverage of the Virgo area with the GUViCS program. *Colored areas* indicate GALEX pointings with various depth. *Circles* indicate GI data obtained especially for the GUViCS proposal in 2010. The *solid line* indicates the NGVS area. The dashed lines indicate the positions of the HeViCS fields

2 Formation of Dwarf Ellipticals by Ram Pressure

Implementing the same ram-pressure as in the previous section to models of spirals/irregulars with intermediate and low-mass, it was found that they (1) loose large quantities of gas (2) do not form stars that would have been formed in the absence of the interaction. This results in galaxies of lower masses, and redder colors corresponding to dE galaxies, such as those observed in Virgo. "Transitional" galaxies (in between star forming and dE) are also reproduced. This suggests that Virgo dE galaxies can be formed by ram-pressure, as detailed in [2, 3].

3 GUViCS and Other On-Going Surveys of Virgo

The Virgo cluster has been observed quite extensively in the UV with the GALEX satellite, especially in the framework of the GUViCS program[1] (Boselli et al., in preparation). The coverage in the GALEX NUV band is shown in Fig. 1. The UV gives us access to events occurring on time scales of a few 100 Myr, much shorter than the timescales derived from the optical. Actually, the combination of multi-wavelength data will allow us to put strong constraints on the ages of ram-pressure (or other) events. This makes GUViCS extremely powerful, especially in combination with other surveys of Virgo, such as the NGVS survey in the optical (Ferrarese et al., in preparation) whose contour is indicated in Fig. 1, ALFALFA in HI, and the HeViCS Herschel program in the far infrared [5] whose contours

[1]The GUViCS web page is http://galex.oamp.fr/guvics/.

are indicated in Fig. 1. This combination of multi-wavelength data will allow us to study in details and for many objects the transformation of galaxies in Virgo under the influence of ram-pressure (similarly to the examples mentioned in Sects. 1 and 2) or other interactions.

References

1. S. Boissier, N. Prantzos, MNRAS **312**, 398 (2000)
2. A. Boselli et al., A&A **489**, 1015 (2008)
3. A. Boselli et al., ApJ **674**, 742 (2008)
4. A. Boselli et al., ApJ **651**, 811 (2006)
5. J.I. Davies et al., A&A **518**, L48 (2010)

An X-Ray View of Polarized Radio Ridges

M. Weżgowiec, D.J. Bomans, M. Ehle, K.T. Chyży, M. Urbanik, and M. Soida

Abstract Observations of group and cluster galaxies reveal a large variety of interactions between galaxies and with the surrounding medium. Radio polarized emission is a very sensitive tracer of such interactions. In particular, regions of strong gas compression are often seen as polarized radio ridges, which mark regions of an enhanced magnetic field. Such enhancements can be however produced either by ram pressure or tidal effects. We present X-ray observations of a perturbed Virgo Cluster galaxy NGC 4569 and show that the analysis of the hot gas distribution and its spectral analysis in the regions where polarized radio ridges are seen, can help to distinguish between the two above scenarios.

Environmental interactions. Cluster and group galaxies seem to experience a variety of environmental effects, which can cause truncations of their HI disks, as well as changes in the global morphology of galaxies (Bos [1]). The most common environmental effects are believed to be ram pressure and tidal effects. They can affect the interstellar medium (ISM), as well as the magnetic field of a galaxy. Indeed, sensitive radio observations show asymmetries and ridges of the polarized

M. Weżgowiec (✉) · D.J. Bomans
Ruhr-Universität Bochum, Universitätsstrasse 150, 44780 Bochum, Germany
e-mail: mawez@astro.rub.de; bomans@astro.rub.de

M. Ehle
ESAC, XMM-Newton Science Operations Centre P.O. Box 78, 28691 Villanueva de la Cañada, Madrid, Spain
e-mail: Matthias.Ehle@sciops.esa.int

M. Weżgowiec · K.T. Chyży · M. Urbanik · M. Soida
Obserwatorium Astronomiczne Uniwersytetu Jagiellońskiego, ul. Orla 171, 30-244 Kraków, Poland
e-mail: markmet@oa.uj.edu.pl; chris@oa.uj.edu.pl; urb@oa.uj.edu.pl; soida@oa.uj.edu.pl

I. Ferreras and A. Pasquali (eds.) *Environment and the Formation of Galaxies: 30 years later*, Astrophysics and Space Science Proceedings,
DOI 10.1007/978-3-642-20285-8_49, © Springer-Verlag Berlin Heidelberg 2011

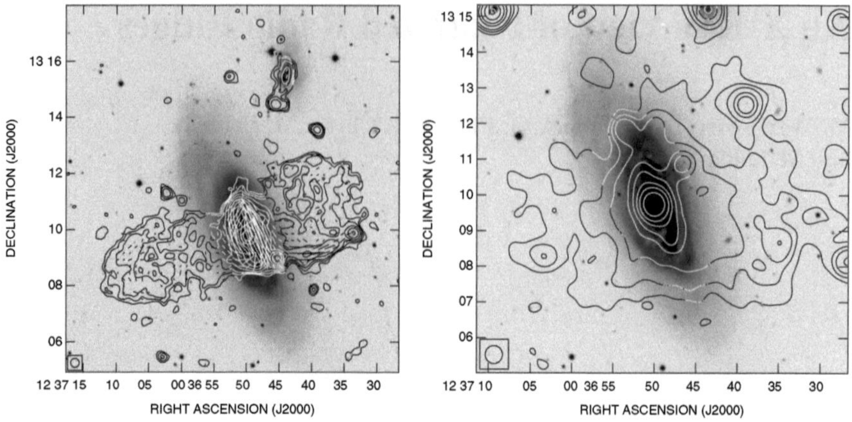

Fig. 1 The DSS blue image of NGC 4569 overlaid with: *left* – contours of total radio intensity of at 4.86 GHz and apparent B-vectors proportional to the polarized intensity, *right* – X-ray contours of soft 0.2–1 keV emission

emission. The analysis of the X-ray emitting hot gas in such galaxies allows us to investigate interaction scenarios for a galaxy [3].

Shock heating. To distinguish between ram pressure or tidal origin of an observed polarized radio ridge, we use the fact that a shock moving supersonically would heat up the gas. Following [4] we adopt a formula for the shock heating, which can be simplified to $\frac{T_2}{T_1} = M$ – a ratio of post- and preshock temperatures equals to the shock Mach number.

NGC 4569 – a case study. The Virgo cluster spiral galaxy NGC 4569 has been studied in the radio domain by [2] (also Chyży et al. in preparation). The galaxy shows impressive radio lobes, unusual for a spiral galaxy (Fig. 1 left). The authors suggest that one possible explanation of the origin of these features could be a strong wind in a past starbursting phase. Previous X-ray studies by [5] showed hints for extraplanar hot gas. We performed sensitive X-ray observations of this galaxy to further investigate this assumption. The polarized radio ridge visible in the south of the western lobe does not have its counterpart in the X-ray map. However, in this region the gas is found to be hotter than in other parts of the western lobe. This could be caused by the galactic wind hitting previously stripped HI gas. As a result the outflowing gas would be compressed and the magnetic field enhanced.

Other examples. We have performed the same analysis with X-ray data of other perturbed spiral galaxies – NGC 4254 in the Virgo Cluster and NGC 2276 in the NGC 2300 group. In the former case a prominent polarized radio ridge in the southern edge of the galactic disk is most likely caused by tidal interation with a close companion. The latter case shows hints for tidal perturbations, however ram pressure effects can not be ruled out. All the mentioned findings will be fully presented in an upcoming paper (Weżgowiec et al. in preparation).

Acknowledgements This work is based on observations obtained with XMM-Newton, an ESA science mission with instruments and contributions directly funded by ESA Member States and NASA. This work is supported by DLR Verbundforschung "Extraterrestrische Physik" at Ruhr-University Bochum through grant 50 OR 0801 and by the Polish Ministry of Science and Higher Education, grant 2693/H03/2006/31.

References

1. A. Boselli, G. Gavazzi, PASP **118**, 517 (2006)
2. K.T. Chyży et al., A&A **447**, 465 (2006)
3. M.E. Machacek, C. Jones, W.R. Forman, ApJ **610**, 183 (2004)
4. J. Rasmussen, T.J. Ponman, J.S. Mulchaey, MNRAS **370**, 453 (2006)
5. D. Tschöke et al., A&A **380**, 40 (2001)

Gravity at Work: How the Build-Up of Environments Shape Galaxy Properties

S. Khochfar

Abstract We present results on the heating of the inter-cluster medium (ICM) by gravitational potential energy from in-falling satellites. We calculate the available excess energy of baryons once they are stripped from their satellite and added to the ICM of the hosting environment. This excess energy is a strong function of environment and we find that it can exceed the contribution from AGNs or supernovae (SN) by up to two orders of magnitude in the densest environments/haloes. Cooling by radiative losses is in general fully compensated by gravitational heating in massive groups and clusters with hot gas temperature >1 keV. The reason for the strong environment dependence is the continued infall of substructure onto dense environments during their formation in contrast to field-like environments. We show that gravitational heating is able to reduce the number of too luminous galaxies in models and to produce model luminosity functions in agreement with observations.

1 Introduction

Within the ΛCDM paradigm of structure formation, the growth of galaxies is governed by dissipational processes such as cooling of gas and associated star formation, as well as dissipationless processes such as mergers of stellar systems that are already in place e.g. [30]. The interplay of such processes regulates the properties of individual galaxies. At early times during the cosmic evolution ($z \geq 2$), when cooling times are short in low mass [10] and massive haloes [5, 17] dissipational processes associated with star formation either is disc-like systems or gas rich mergers [19] dominate the growth of galaxies. With the onset of downsizing

S. Khochfar (✉)
Max-Planck-Institute for Extraterrestrial Physics, Giessenbachstrasse 1, 85748 Garching, Germany
e-mail: sadeghk@mpe.mpg.de

I. Ferreras and A. Pasquali (eds.) *Environment and the Formation of Galaxies: 30 years later*, Astrophysics and Space Science Proceedings,
DOI 10.1007/978-3-642-20285-8_50, © Springer-Verlag Berlin Heidelberg 2011

in the star formation rate of galaxies at low redshifts e.g. [24] this picture changes, in particular for massive early-type galaxies that can accrete up to 80% of their stellar mass by $z = 0$ from satellites [16]. The importance of such 'dry' mergers for the growth of massive galaxies at late times has been investigated e.g. in [11, 18, 22] and recently linked to the size-evolution of massive early-type galaxies [15, 23, 27].

Feedback is generally assumed to play an important role in trends as the ones mentioned above e.g. [25]. In particular supernovae feedback has been considered in shaping the faint-end of the luminosity function [1, 20]. At the luminous end the situation appears more complicated with several competing effects at play as e.g. AGN-feedback [3] or gravitational heating by in-falling sub-structure [9, 14].

The observed properties of the galaxy population in high-density environments show distinct properties in contrast to the field galaxy population. The population of spiral galaxies is gas-poor [21] and has smaller HI discs [28] compared to spiral galaxies in the field. Besides affecting the properties of individual galaxies, the morphological mix of the population as a whole changes in high density environments as well. The so-called density-morphology relation shows an increasing trend in the fraction of early-type galaxies as a function local density [6]. The process of ram-pressure stripping [8] is able to remove gas from satellite galaxies orbiting in high density environments, and truncate disc sizes. The change in morphological fraction is generally attribute to a higher fraction of mergers during the formation epoch of the high density environment.

While the above mentioned processes *actively* change individual galaxies we here address how the build-up of the environment *passively* affects the state of the central galaxy population living in it via the ability/inability to cool gas. Numerical simulations show that shock heating of gas falling into the cluster potential is able to heat gas initially to the virial temperature of the cluster [7]. In addition conversion of gravitational potential energy of orbiting satellites is able to heat the inter-cluster medium (ICM) [4, 14]. In the following we will focus on the latter effect and include it into a semi-analytic model (SAM) to estimate its impact on the galaxy population.

2 Model

The results we present here are based on the SAM described in detail in [12, 14], using the three-year WMAP cosmology with $\Omega_0 = 0.27$, $\Omega_\Lambda = 0.73$, $\Omega_b/\Omega_0 = 0.17$, $\sigma_8 = 0.77$, $h = 0.71$ [26]. We briefly summarize here the main physical models that are important in high density environments, and that we have included in our SAM.

2.1 Shock Heating

Shock-heating of satellite gas takes place when a small satellite falls within the large potential well of a host dark matter halo and starts interacting with the hot halo gas. Generally SAMs assume that the shock heating is taking place instantaneously, thereby removing all hot halo gas from satellites and adding it to the hot gas reservoir of the central galaxy [10]. It has been that this process, if modeled instantaneously might be too efficient resulting in too many passive satellite galaxies in high density environments. [29]. We adopt a simple prescription for the process of shock heating assuming that the rate at which mass is shock heated from the satellite is related to the halo dynamical time t_{dyn} via $\dot{M} \sim M/t_{dyn}$. This is effects allows the satellite to hold on to his hot gas for a longer period of time and to continue cooling.

2.2 Ram Pressure Stripping

Following [8], we assume that gas is stripped from the satellite once the dynamical pressure is able to over- come the gravitational force binding the gas to the satellite. In terms of energy deposited within the satellite gas, this can be approximated by

$$\dot{E}_{ram} = \rho_{hot} v_{\perp}^{3} \pi r_{h}^{2} \tag{1}$$

Here, we take v_{bot} as the velocity of the satellite perpendicular to its disc orientation and assume that the orbital velocity of the satellite is comparable to the sound speed c_{gas} of the hot gas. The efficiency of this process for a face-on disk should be maximal, and the efficiency for an edge-on disk should be minimal; we take this into account by assuming that the disk orientation is random with respect to the infall direction. This is in good agreement with cosmological dark matter simulations that show that the spin vectors of merging dark matter halos are randomly aligned to each other [13]. For simplicity, we calculate the density ρ_{hot} by taking the average density of hot gas within the host halos virial radius, and take r_h to be the characteristic half-mass radius of the gas within the satellite. The rate of stripped material is then calculated dividing $\dot{E}_{ram}$ by the specific energy of the gas on which it is acting.

2.3 Gravitational Heating

The energy needed to strip baryonic material from satellites is provided by the conversion of gravitational potential energy, that the satellite gains during its infall on a slightly bound $E_{bind} \sim 0$. Parabolic orbits have indeed been shown to be the most probable based on cosmological N-body simulations [13]. Gas that is stripped from a satellites can in principle have potential energy left, after it has been used up for the stripping process, and we assume here that this energy will be used to heat the inter-cluster medium. In practice we apply following equation to calculate the amount of gravitational potential energy left to heat the ICM:

$$\dot{E}_{grav} = \sum_{i=1,n_{sat}} \dot{M}_{gas,i} \left(\Delta\phi - \Delta E_{strip} \right). \tag{2}$$

The amount of stripped gas for each satellite i orbiting within the host halo is $\dot{M}_{gas,i}$ and the total energy needed to strip the material from the satellites is ΔE_{strip}.

The heating of the ICM based on (2) predicts that for the same sum of stripped material from massive satellites, heating is less efficient than if that material would be stripped from less massive satellites.

3 Results

The time integral of (2) over the evolution of a high density environment can be quite substantial and, if expressed in terms of $E_{grav,tot} = \epsilon_{grav} m_* c^2$, we find values for ϵ_{grav} ranging from a few time 10^{-8} to a few times 10^{-4} in galaxies with stellar masses $M_* \sim 10^{10}$ and $\sim 5 \times 10^{12}$ $M_\odot$, respectively. To emphasize that this trend is driven by the environment, we show in Fig. 1 the trend with halo mass which we take as a proxy for the environment. Massive haloes have more gravitational heating during their evolution than less massive ones. One can

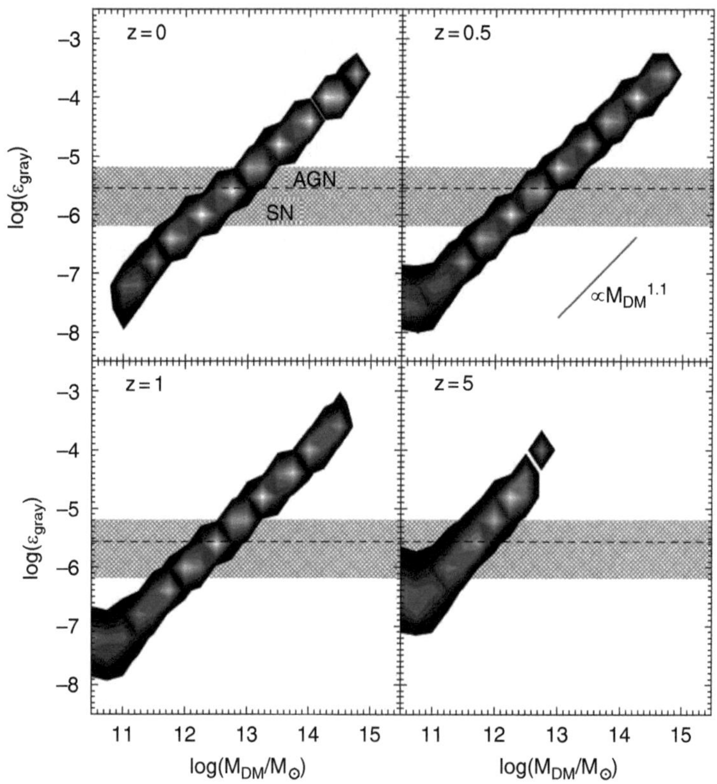

Fig. 1 The integrated efficiency of gravitational heating ϵ_{grav} as a function of host halo mass. Galaxies living in the most massive haloes always experience the largest contribution from gravitational heating, due to the continuous infall of satellites. *Shaded areas* show the expected range in efficiencies for SN and AGN-feedback

understand this behavior by considering the accretion of satellites onto halos. In general, the most massive halos at any redshift had the largest accretion rates in the past, which explains the large amount of gravitational heating and its dependence on halo mass. It is worth comparing gravitational heating to other common heating mechanisms such as supernovae and AGNs. Comparing ϵ_{grav} to $\epsilon_{SN} \sim 2.8 \times 10^{-6}$ and $\epsilon BH \sim 6.5 \times 10^{-6} - 6.5 \times 10^{-7}$ shows that in general, gravitational heating is more efficient than supernova feedback only in galaxies larger than a few times 10^{11} $M_{\odot}$ and in halos more massive than 5×10^{12} $M_{\odot}$ at $z = 0$. This regime corresponds to massive field galaxies and extends into group-like environments. For even more massive galaxies and dark halo masses larger than 10^{13} $M_{\odot}$, gravitational heating starts dominating over proposed AGN feedback rates.

To illustrate the contribution from gravitational heating, we display the contours of the conditional probability for the ratio of heating to cooling that individual galaxies experience in a given environment. The top panel in Fig. 2 shows the

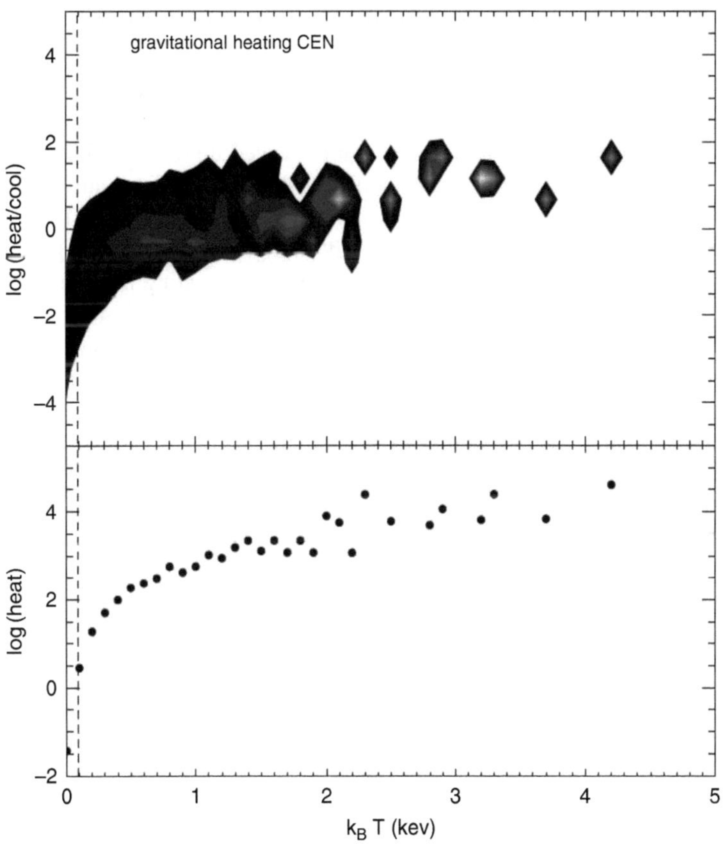

Fig. 2 *Top*: the probability distribution for the ratio of heating to cooling rates as function of environment. The x-axis shows the temperature of the hot halo gas as a proxy for environment. *Bottom*: The average gravitational heating rate as a function of environment

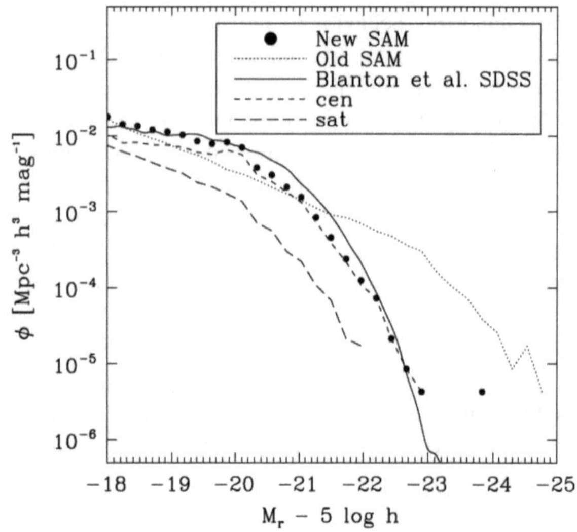

Fig. 3 The r-band luminosity function of galaxies in our model including gravitational heating (*filled circles*) and without (*dotted line*), compared to the SDSS results (*solid line*) by [2]

probability contours for central galaxies at $z = 0.1$ in our simulations. We translate the deposited energy per unit time into a heating rate, labeled *heat* in Fig. 2, by calculating the amount of cold gas that can be heated up to the virial temperature of the dark matter halo in which the galaxy resides. The cooling rate, labeled *cool* is the standard radiative cooling rate that we calculate in our SAM based on the prescription in [12]. Figure 2 shows clearly that gravitational heating dominates in environments with temperatures above ~keV. The heating rate for the central galaxies can be up to 10^2 times larger than the cooling rate, and in the most dense environments the heating rate becomes 10^4 $M_\odot$ yr^{-1}. The heating rate shows a clear environmental dependence that reflects the higher abundance of satellites that contribute to gravitational heating. From these results, one expects that star formation in central galaxies of dense environments will be shut down.

To further illustrate the effect of gravitational heating on the galaxy population we show in Fig. 3 the luminosity function of galaxies at $z \sim 0$. We find a good match to the observed one by [2]. The effect of gravitational heating is strongest for massive galaxies. These tend to live predominantly in high density environments, subject to gravitational heating. For comparison we also show a model without the contribution of gravitational heating. This model severely overproduces the number density of luminous galaxies.

4 Conclusions

We here show the effects that the surplus of gravitational potential energy from infalling baryonic matter can have on the heating of the ICM. Galaxies in low density environments are generally not affected by gravitational heating as the contribution is small compared to other feedback sources like SN or AGN feedback. We find that in the most dense environments, like massive groups or clusters, the constant infall of substructure is able to provide enough energy to level, and even exceed the losses due to radiative cooling, hence stopping any cooling in such environments. The integrated contribution from gravitational heating is in cluster environments clearly larger than the contribution from AGNs suggesting that this could be an important feedback source. This conclusion is supported by looking at the luminosity function of galaxies. Including gravitational heating is able to reduce the amount of too luminous galaxies and to bring them in agreement with observations. To further test the importance of gravitational heating high-resolution simulations of galaxy clusters are required. Such simulations should be able to reveal self-consistently the importance of the effects discussed here.

Acknowledgements SK would like to thank Anna Pasquali and Ignacio Ferreras for organizing such a stimulating meeting at the JENAM 2010.

References

1. A.J. Benson et al., ApJ **599**, 38 (2003)
2. M.R. Blanton et al., ApJ **592**, 819 (2003)
3. R.G. Bower et al., MNRAS **370**, 645 (2006)
4. A. Dekel, Y. Birnboim, MNRAS **383**, 119 (2008)
5. A. Dekel et al., Nature **457**, 451 (2009)
6. A. Dressler, ApJ **236**, 351 (1980)
7. C.S. Frenk et al., ApJ **525**, 554 (1999)
8. J.E. Gunn, J.R. Gott III, ApJ **176**, 1 (1972)
9. P.H. Johansson, T. Naab, J.P. Ostriker, ApJ Lett. **697**, L38 (2009)
10. G. Kauffmann, S.D.M. White, B. Guiderdoni, MNRAS **264**, 201 (1993)
11. S. Khochfar, A. Burkert, ApJ Lett. **597**, L117 (2003)
12. S. Khochfar, A. Burkert, MNRAS **359**, 1379 (2005)
13. S. Khochfar, A. Burkert, A&A **445**, 403 (2006)
14. S. Khochfar, J.P. Ostriker, ApJ **680**, 54 (2008)
15. S. Khochfar, J. Silk, ApJ Lett. **648**, L21 (2006)
16. S. Khochfar, J. Silk, MNRAS **370**, 902 (2006)
17. S. Khochfar, J. Silk, ApJ Lett. **700**, L21 (2009)
18. S. Khochfar, J. Silk, MNRAS **397**, 506 (2009)
19. S. Khochfar, J. Silk, MNRAS **410**, L42 (2011)
20. S. Khochfar et al., ApJ Lett. **668**, L115 (2007)
21. S.M. Moran et al., ApJ Lett. **641**, L97 (2006)
22. T. Naab, S. Khochfar, A. Burkert, ApJ Lett. **636**, L81 (2006)
23. T. Naab, P.H. Johansson, J.P. Ostriker, ApJ Lett. **699**, L178 (2009)
24. K.G. Noeske et al., ApJ Lett. **660**, L43 (2007)
25. J. Schaye et al., MNRAS **402**, 1536 (2010)
26. D.N. Spergel et al., ApJ Sup. Ser. **170**, 377 (2007)
27. I. Trujillo et al., ApJ **650**, 18 (2006)
28. J.H. van Gorkom, in *Clusters of Galaxies: Probes of Cosmological Structure and Galaxy Evolution*, Carnegie Observatories Centennial Symposia, Cambridge U.P., p. 305 (2004)
29. S.M. Weinmann et al., MNRAS **394**, 1213 (2009)
30. S.D.M. White, C.S. Frenk, ApJ **379**, 52 (1991)

MIX
Papier aus verantwortungsvollen Quellen
Paper from responsible sources
FSC® C105338

If you have any concerns about our products,
you can contact us on
ProductSafety@springernature.com

In case Publisher is established outside the EU,
the EU authorized representative is:
Springer Nature Customer Service Center GmbH
Europaplatz 3, 69115 Heidelberg, Germany

Printed by Libri Plureos GmbH
in Hamburg, Germany